The Institute of Biology's
Studies in Biology no. 80

Colonization of Industrial Wasteland

Raymond P. Gemmell
B.Sc., Ph.D., F.L.S., M.I.Biol.
Ecologist, Joint Reclamation Team, Greater Manchester and Lancashire County Councils

Edward Arnold

First published 1977
by Edward Arnold (Publishers) Limited
25 Hill Street, London, W1X 8LL

Boards edition ISBN: 0 7131 2586 1
Paper edition ISBN: 0 7131 2587 X

Printed in Great Britain by
The Camelot Press Ltd, Southampton

General Preface to the Series

It is no longer possible for one textbook to cover the whole field of Biology and to remain sufficiently up to date. At the same time teachers and students at school, college or university need to keep abreast of recent trends and know where the most significant developments are taking place.

To meet the need for this progressive approach the Institute of Biology has for some years sponsored this series of booklets dealing with subjects specially selected by a panel of editors. The enthusiastic acceptance of the series by teachers and students at school, college and university shows the usefulness of the books in providing a clear and up-to-date coverage of topics, particularly in areas of research and changing views.

Among features of the series are the attention given to methods, the inclusion of a selected list of books for further reading and, wherever possible, suggestions for practical work.

Readers' comments will be welcomed by the author or the Education Officer of the Institute.

1977

The Institute of Biology,
41 Queen's Gate,
London, SW7 5HU

Preface

In Britain, the exploitation of natural resources for energy, minerals and manufacturing materials has created nearly one hundred thousand hectares of industrial wasteland. Throughout the world, industrialized countries have experienced similar environmental despoliation and the problems which lie in its wake. Not until recent times, however, has proper attention been given to the tasks of restoration and rehabilitation of this spoiled land.

In recent years there has been an upsurge of interest in the ecology of industrial wasteland and the problems of vegetation establishment which have to be solved before it can be effectively restored. It is a pity that no real attempt has been made to summarize the available information in a single text because the subject is highly topical and is of great interest to the student of environmental sciences. I hope that this booklet will help to overcome this omission and give the reader an insight into a particularly rewarding field of study which is of great practical application in our present industrial society.

Preston, 1976

R. P. G.

Contents

1 Industrial Wasteland and Restoration

1.1 Creation of industrial wasteland in Britain

The natural environment has suffered great damage as a result of man's industrial activities. One of the most obvious examples of disturbance to the ecosystem is the obliteration of plant and animal communities by the deposition of wastes generated from mining and other industrial operations. The fact that such wastes are often toxic and infertile has resulted in the creation of an environment which is not only hostile to recolonization by living organisms but is unsightly, polluting, and generally depressing in its effect on the local populace and communities.

Most of the wasteland in Britain has resulted from industries established during the Industrial Revolution which are now in a state of decline or have disappeared completely. The most extensive damage over the country as a whole has been inflicted by the deposition of spoil from the coal mining industry, the worst affected areas being in the coalfields of North England, the North Midlands, South Wales and parts of Central Scotland. Roughly half of the material extracted underground in deep coal mining ends up as deposited waste. At Leigh, in the heart of the old South Lancashire coalfield, this has resulted in the accumulation of an immense tip covering an area of 150 hectares which surrounds and almost isolates the mining village of Gin Pit. The barren, ugly, and now useless landscape of black, acidic spoil, overshadows the more recent housing developments in its neighbourhood.

Of the metalliferous industries which have resulted in spoiled land, the extraction and smelting of iron has had the greatest impact. Although the Ironstone Restoration Fund has ensured that land affected by iron ore extraction has been restored in recent years, the smelting of iron ore in blast furnaces throughout the country has left its mark on the landscape. The old blast furnaces have closed down, but have left behind them huge tips of waste known as blast furnace slag. The town of Dowlais, which lies at the head of Merthyr Vale in South Wales, is overlooked by what is probably the largest and ugliest waste tip from iron smelting in the country.

The extraction and smelting of non-ferrous metal ores have now practically ceased in Britain. Whereas colliery waste heaps and blast furnace slag tips are usually extensive and often located in or near urban areas, non-ferrous metal mine spoil heaps are relatively small and dispersed over the countryside, often in remote areas of beautiful scenery. The smelters, on the other hand, were situated near coal supplies and concentrated into two or three places only, the most important centre being the Lower Swansea Valley. At one time Swansea was a world centre for non-ferrous metal smelting; the industry later declined and was replaced by iron and steel smelting. Today, the Lower Swansea Valley is the largest single area of industrial wasteland in Britain, being choked with almost ten million tonnes of toxic and infertile wastes which repel would-be developers and are overlooked by almost one-third of Swansea Borough's inhabitants.

In some parts of Britain, particularly in North-West England around Manchester and the Mersey Belt, toxic wastes have been deposited from the chemicals industry. This kind of dereliction is difficult to restore because the wastes are often highly toxic to plants and may cause water pollution. One such tip near Bolton consisted of nearly 750 thousand tonnes of toxic chromate waste; the site has now been successfully reclaimed to attractive amenity grassland and playing fields.

Although terrestrial pollution by the coal-mining, metalliferous and chemical industries has declined, derelict land is still being created by deposition of different types of waste. In recent years the disposal of ash from power stations burning pulverized coal has become a major problem. Some uses have been found for this waste in the building industry but a vast quantity is tipped on land.

The extraction of china clay in Cornwall generates nine tonnes of waste for every tonne of china clay produced. The residues are deposited in pits, lagoons, or on surface tips. The development of this industry has now had a marked effect on the landscape of the South-West; the vast conical tips of whitish material are notable features in the countryside of the area.

Another relatively new industry causing increasing problems is the extraction of fluorspar in Derbyshire. The waste is disposed of in lagoons and spoil heaps and is causing increasing concern because the industry is situated in the Peak District National Park.

Wastes contributing to dereliction on a smaller scale include materials from sand and gravel workings, stone and slate quarries, by-product calcium sulphate, domestic refuse and incinerator ash. Table 1 gives a summary of the locations and disposal of the most important industrial waste materials generated in Great Britain.

It will be noticed that solid materials such as domestic refuse, sewage wastes, agricultural wastes, and various types of liquid and solid industrial wastes are excluded from Table 1. The disposal of these materials poses rather special and different problems for they are collected and tipped by local authorities or contractors and then 'blinded' with cover material. The disposal sites are generally controlled by planning permissions and covered by recommendations of the Department of the Environment so as to ensure satisfactory after-treatment and prevention of pollution during tipping. In many cases, however, the recommendations are not followed very closely and

Table 1 Production and location of wastes in Great Britain. (Adapted from GUTT et al., 1974. *A Survey of the Locations, Disposal and Prospective Uses of the Major Industrial By-products and Waste Materials*, Current Paper CP 19/74, Building Research Establishment, Watford.)

Type of waste	*Stockpile M tonnes*	*Area ha*	*Production M tonnes/year*	*Principal locations*	*Method of disposal*
Colliery spoil	3000	15 000	50	Coalfields of N. and NE. England, Midlands, S. Wales, C. Scotland and Kent	Mainly tipped as waste, some as fill and in manufacture of building materials
Blast furnace slag	Not quantified but small		9	S. Wales, Corby, Sheffield, Scunthorpe, Consett, Teesside, Glasgow, Cumbria	Current production all used, mainly in roads—some old tips remain
Steel-making slag	Not quantified but small		4	As above for blast furnace slag	2 M tonnes returned to blast furnaces. Some used as roadstone. Rest tipped or used as fill around steelworks
Metal mine spoils	Not known	Not known	0.46 (Sn only)	Cornwall, N. and C. Wales, Derbyshire, Lake District, Pennines	Tips scattered in countryside
Metal smelter wastes	Not known	Not known	0.24	Swansea (Zn, Cu), Widnes and Walsall (Cu), Bristol (Zn, Pb), Hull and Liverpool (Sn)	Major utilization, old tips remain
*Power station ash	Not known	Not known	10	Countrywide	Lagooned and tipped, about 60% used in construction
Slate waste	Over 300	Not known	1.2	Wales, Lake District, Cornwall, Devon and Scotland	Most tipped
China clay waste	280	800	22	Cornwall, Devon	Most tipped, minor usage for building
Quarry wastes	Not known	Not known	Not known	Countrywide	Minor usage for building, rest tipped
Fluorspar mine waste	Not known	Not known	0.23	Pennine areas of Derbyshire and Durham	Tailings lagoons
Calcium sulphate	Not quantified but small		2.1	Produced in various locations from phosphoric and hydrofluoric acid plants	Mainly disposed of at sea, small amount tipped on land
*Furnace clinker	None	None	1.4	Countrywide	All now used for building
*Incinerator ash	Not known	Not known	0.6	Countrywide	Most is tipped
Chemical wastes	Not known	Not known	Not known	Mersey Belt and Cheshire (alkali and salt wastes), S. Wales and Fife (red mud from alumina manufacture)	Tipped or lagooned

Items marked * refer to England and Wales.

sometimes not at all. It is hoped that this situation will improve with the passing of more effective legislation and better control by the local authorities concerned.

1.2 Overseas problems

Many of the wasteland problems discussed in the previous section are worldwide in scope, being faced by most highly industrialized nations. Some countries, however, have to deal with rather special and unusual kinds of land reclamation. In the United States, the extraction of coal by strip mining has caused immense devastation and acid drainage pollution. On the Witwatersrand in South Africa, dumps of spoil from gold mining, sometimes several square miles in extent, pose health hazards because of fine siliceous dusts and cause pollution of streams due to run-off. In Malaya and Nigeria, areas have been devastated by tin mining and its associated spoil dumps. Non-ferrous metal mining has practically ceased in Britain but several countries have large, open-cast copper and other metal mines.

1.3 Future problems

The extraction of oil-shale to supplement petroleum resources, particularly in the United States and Canada, becomes more likely as a major future development. This will generate very substantial waste disposal problems for, in most cases, from 80 to 90% of the original rock appears as spent shale.

The possibility of large-scale open-cast mining of metalliferous ores such as copper and zinc cannot be discounted in Britain. Since most of the deposits lie in National Parks, any mining operations and disposal of spoil would be a grave threat to the countryside.

1.4 Social problems and the need for colonization

The complete restoration of wasteland as well as the future protection of natural resources in Britain and indeed in many countries is now recognized as a problem of great importance. It is accepted that wasteland imposes an economic expense on areas where it is prevalent, primarily because it absorbs scarce land and creates an unfavourable environment which, by deterring industrial and housing development in the surrounding areas, can depress economic opportunity. It has an offensive and depressing effect on local communities and may even be the cause of substantial outward migration. For these reasons, restoration has been seen as a vital part of policy aimed at rejuvenation of major industrial areas of the past and an effective way in which the living conditions of their present inhabitants can be improved.

The most obvious effects of tipping waste on land are that the landform is altered and woodlands, fields and soils are buried beneath mountains of refuse. Major features of affected land are steep slopes, mechanical instability, erosion, disruption of natural drainage systems, and absence of vegetation due to infertility and/or toxicity. Thus the land cannot be used for industrial or housing development, agriculture, forestry, amenity, recreation, or any other such purpose unless it is first subjected to special reclamation treatment (Fig. 1–1).

Fig. 1–1 Chemical waste tip. The two dark areas on the right of the picture consist of iron pyrite and ashes. The white areas are waste lime and salt residues. In the foreground limestone blocks can be seen which have been deposited to neutralize the acid drainage from the pyritic wastes.

A further problem is that wastes may pollute water and the air. Wind-blown dusts from unstable metalliferous tips may pose health hazards for man, either directly so or via livestock and crops. Acid drainage from colliery spoil has severely damaged the biota of natural water courses in many countries and toxic metals from metalliferous wastes have had similar effects. Chemical wastes may release extremely toxic effluents. Table 2 lists some of the pollution problems caused by industrial wastes.

1.5 Landscape restoration in Britain

The official definition of derelict land for the purpose of local authority

Table 2 Pollution problems caused by wastes.

Type of waste	*Nature of pollution*
Colliery waste	1 Acid run-off and seepage into natural watercourses 2 Emission of suspended solids in run-off 3 Deposition of ferruginous compounds in stream and river beds 4 Emission of wind-blown dusts due to surface instability 5 Emission of noxious gases due to spontaneous combustion 6 Mechanical instability leading to landslip
Metalliferous smelter wastes and mine spoils	1 Source of water-borne toxic metal contaminants and acidity 2 Source of wind-blown dusts laden with toxic metals and causing a health hazard 3 Source of high levels of toxic metals in plants causing a danger to grazing livestock
Chemical wastes	1 Seepage of highly toxic chromates into watercourses from chromate waste tips 2 Release of sulphides from alkali tips causing pollution of watercourses 3 Release of acidity from pyritic waste tips 4 Release of effluents containing various toxic salts which contaminate watercourses 5 Release of suspended solids into water
China clay waste	1 Contamination of watercourses by suspended solids

surveys and the payment of government grants for restoration is 'land so damaged by industrial or other development that it is incapable of beneficial use without treatment'. This definition excludes land subject to planning permissions requiring restoration, land still in use, and urban sites awaiting development.

1.5.1 Government legislation and costs

The British Government has several Acts of Parliament which provide local authorities with generous grants to restore derelict land. For instance, there are the 1966 Local Government and Industrial Development Acts and 1970 Local Employment Act which provide 50 to 85% of the cost of clearance or treatment in appropriate cases. Until very recently the Exchequer Grant available has been 75% generally and 85% in development areas. Now, however, the Government provides a 100% grant for reclaiming derelict land in approved cases. This is calculated on the basis of net cost:

cost of land acquisition *plus* cost of restoration *less* after value

Land costs account for about a third of the cost of restoration because it is often necessary to acquire much land surrounding waste tip sites so that the material can be spread out for landscape modelling. Allowing for the cost of land acquisition, the usual cost of restoration varies from £5000 to £10 000 per hectare at 1975 prices.

1.5.2 Restoration in North-West England

The Civic Trust has estimated that there are approximately 100 000 hectares of industrial wasteland in Britain. Two-thirds of the land needing treatment in England is situated north of the Trent, the worst affected county being the Metropolitan County of Greater Manchester. In some areas east of Wigan up to 25% of the total land area is derelict. According to the report of the North West Joint Planning Team in July 1973 'Strategic Plan for the North West', the North-West region as a whole is the worst in the country for the incidence of derelict land. The report also concluded that 'assuming resources for tackling environmental problems are limited, the most cost-effective programmes of pollution control in terms of type and distribution of benefits are for the reduction of air pollution and the reclamation of derelict land'.

Surveys in the North-West of England during the period 1964 to 1974 indicate that the amount of derelict land has increased at a rate of about 280 hectares per annum, taking the period as a whole. It appears that much of the apparent net increase of 120 hectares per year is derived from re-surveys being more critical. Even so, more dereliction is being brought about through closure of mineral workings and there is an extensive potential dereliction because of current and proposed workings and waste tipping which are not subject to adequate restoration conditions.

Although the Government set a target for clearance of the worst dereliction by 1980, the 'Strategic Plan for the North West' concluded that, assuming 320 hectares were cleared in 1973 and a similar rate each year, the existing dereliction would not be cleared until 1994.

1.5.3 Technical problems of restoration

One of the first considerations in carrying out a land reclamation scheme is to define the eventual land-use of the area. This may be residential or industrial development, agriculture, forestry, or land for amenity such as public open space, recreation, or wildlife habitats. Normally, the after-uses are considered within a proper planning framework and in accordance with a County Development or Structure Plan. This ensures that a double benefit is obtained: an eyesore or pollution source is removed and a new land use is provided.

There are necessarily engineering problems to be tackled:

1. *Drainage* mining subsidence and tipping cause disruption of natural drainage systems. The waste may be impermeable and so adequate

ditches and drainage facilities are required. Balancing ponds may be needed to control the discharge of site run-off.

2. *Slopes* steep slopes and instability are common. The waste heaps must be surveyed in detail and the amounts of cut and fill calculated so that the final landform can be achieved with a minimal shifting of material.
3. *Fire* shale tips may burn so it is necessary to ensure compaction in order to exclude air.
4. *Shafts* old pit shafts must be filled in or sealed.
5. *Site materials* these have to be assessed in terms of stability, likelihood of settlement after spreading, and suitability for being built on.

Finally, the ecological and planting problems of the wastes have to be considered if vegetation is planned. This problem was recognized in Lancashire in the early 1950s when reclamation in Britain was in its infancy. An experiment was set up by Lancashire County Council at Bickershaw, the raw shale being given different treatments of lime and fertilizer without topsoil. The results were very encouraging, growth similar to that in the surrounding fields being achieved. The area cannot now be identified as an old tip site. Other trials produced good stands of trees in colliery shale. However, many of the larger sites presently being reclaimed are much more difficult and it has been found to be extremely important to carry out a programme of after-management otherwise the vegetation may deteriorate or die back completely.

1.6 Chemistry and origin of wastes

The majority of materials occurring on waste land are of mineralogical origin. Wastes such as colliery spoil which are dumped in their natural state develop toxicity as a result of chemical and physical changes induced by weathering. On the other hand, smelter and chemical wastes are initially hostile to growth because of the formation of toxic ingredients by chemical reactions promoted by heating in the presence of additives.

(*a*) *Colliery spoil* Colliery spoil consists mainly of material formed during the Upper Carboniferous era some 250 million years ago. The coal-bearing strata are associated with deposits of shale, sandstone, limestone and clay which are exposed during mining operations. Table 3 shows the principal minerals present though these may vary to a very great extent depending on the conditions prevailing at the time the deposits were formed.

Iron pyrite, which is the main toxic ingredient of many colliery shales, is derived from the association of organic and mineral sediments. The former were subjected to partial decomposition which resulted in reducing conditions being formed causing sulphates to be reduced by

Table 3 Mineral constituents of colliery spoil. (After DOUBLEDAY, G. P., 1971, *Landscape Reclamation*, **1**, IPC Business Press Ltd; Guildford, Surrey.)

Mineral	*Chemical composition*
Clay minerals	Aluminosilicates
Quartz	SiO_2
Orthoclase felspar	$KAlSi_3O_8$
Haematite	Fe_2O_3
Magnetite	Fe_3O_4
Goethite	$HFeO_2$
Iron pyrite	FeS_2
Siderite	$FeCO_3$
Jarosite	$KFe_3(SO_4)_2(OH)_6$
Tourmaline	Aluminium borosilicate containing alkali metals or Fe and Mg
Soluble salts	Chlorides of Na, K, and Mg, etc. Sulphates of Mg and Ca, etc.

bacterial action to sulphides and ultimately pyrites (FeS_2). When the shales are exposed by mining the reverse reaction takes place causing acidity and other problems.

(*b*) *Blast furnace slag* Iron is obtained by the chemical reduction of oxide or carbonate ores (Fe_2O_3, Fe_3O_4 and $FeCO_3$). In the old types of blast furnace the ores were heated in a current of carbon monoxide generated from burning coke, this gas reducing the ores to molten iron. In order to remove the clay and silica impurities in the ores, lime was added which reacted with the alumino-silicates or clays to produce calcium aluminium silicate. This formed as a fused slag on top of the molten iron and was tapped off as waste. For every ten tonnes of iron produced there were between five and eight tonnes of slag from the old blast furnaces. After tipping, the waste was highly alkaline due to the presence of calcium hydroxide and the hydrolysis of basic silicates.

(*c*) *Power station ash* In modern coal-burning power stations the coal is fed to the boilers as a fine powder of pulverized fuel. This burns to produce an ash consisting of tiny glassy spheres which is subsequently treated with water and removed as a slurry for settlement in lagoons. The lagooning process has the effect of removing many of the soluble toxic salts which greatly facilitates colonization. The final chemical composition of a typical ash is shown in Table 4 but, in addition, there may be small amounts of borate which often cause toxicity.

Table 4 Chemical analysis of power station ash. (Adapted from REES, W. J. and SIDRAK, G. H., 1956, *Plant & Soil*, **8**, 141–57.)

Constituent	*% present*
Silica (SiO_2)	43–52
Alumina (Al_2O_3)	16–29
Ferric oxide (Fe_2O_3)	18–20
Ferrous oxide (FeO)	6
Calcium oxide (CaO)	6–7
Magnesium oxide (MgO)	1–2
Manganous oxide (MnO)	0–0.4
Nickel oxide (NiO)	0–0.001
Titanium oxide (TiO)	0.5–1
Alkali metal oxides (NaO and K_2O)	2–4
Carbon (C)	2–4
Sulphur trioxide (SO_3)	1–2
pH	8.5 plus

(*d*) *Extraction and smelting of non-ferrous metal ores* Zinc and copper are mainly extracted from sulphide ores. Due to inefficient separation of the ores from the associated strata, the spoil heaps are contaminated with metallic sulphides which undergo oxidation to release toxic metal cations in solution.

The subsequent smelting of zinc was carried out in two stages which involved roasting the ore in air to form the oxide:

$$2ZnS + 3O_2 \rightarrow 2ZnO + 2SO_2$$

The oxide was then heated with coal or coke in retorts, zinc being distilled off:

$$ZnO + C \rightarrow Zn + CO$$

The waste contains ashes and residual zinc sulphides.

Copper smelting was also conducted in two stages. First, the ore was heated to produce cuprous and ferrous sulphides:

$$2CuFeS_2 + O_2 \rightarrow Cu_2S + 2FeS + SO_2$$

Second, the sulphides were mixed with silica and fused in a converter:

$$2FeS + 2SiO_2 + 3O_2 \rightarrow 2FeSiO_3 + 2SO_2$$
$$2Cu_2S + 3O_2 \rightarrow 2Cu_2O + 2SO_2$$
$$Cu_2S + 2Cu_2O \rightarrow 6Cu + SO_2$$

The slag consists of ferrous silicate ($FeSiO_3$) plus residual copper and iron sulphides.

Oxidation of the zinc, copper, and other metallic sulphides in smelter wastes releases the toxic metal cations in solution.

(*e*) *Chemical wastes* Until recent years, the waste from the obsolete Leblanc Process for sodium carbonate production was a feature of the landscape in parts of South Lancashire. The process involved heating a mixture of sodium sulphate, limestone, and powdered coal. The following reactions occurred:

$$Na_2SO_4 + 2C \rightarrow Na_2S + 2CO_2$$
$$Na_2S + CaCO_3 \rightarrow CaS + Na_2CO_3$$

The residue was extracted with water to remove sodium carbonate. Calcium sulphide was left in the waste with large amounts of calcium hydroxide formed by hydrolysis of the sulphide and residual calcium oxide from excess limestone used in the process. This rendered the waste highly alkaline.

Chromates were obtained from the smelting of chromite ($FeOCr_2O_3$) with sodium carbonate and lime but lime is omitted from the process now. Calcium chromate as well as sodium chromate was formed when lime was used. The calcium salt is less soluble than sodium chromate and so considerable amounts of toxic chromate were left in the waste.

(*f*) *China clay waste* China clay is composed largely of kaolinite, a hydrated aluminium silicate. The china clay is obtained by washing out the associated kaolinized granite with jets of water, the resulting slurry being subjected to separation and dewatering which remove sand, waste rock, overburden, and a micaceous residue. Chemical analysis of the micaceous material shows that it consists of silica (50%), alumina (33%), ferric oxide (2%) and potassium compounds (5%). Compared with the previously discussed wastes, this material is relatively innocuous as far as plant growth is concerned for it contains no heavy metals or oxidizable sulphides.

2 Deficiency of Nutrients

2.1 Nutrient status of soils and wastes

Of the essential nutrients required by plants from soil, nitrogen, phosphorus, potassium, calcium, magnesium and sulphur are taken up in substantial quantities. These are known as major nutrients. The other essential elements are iron, manganese, boron, copper, zinc and molybdenum which, because they are taken up in trace amounts, are termed minor nutrients.

In Britain, the commonest factor limiting the growth of plants in soil is nitrogen supply. Indeed, this situation prevails in many countries throughout the world, particularly where soils are subjected to substantial natural leaching. The second most important nutritional factor restricting growth is phosphate status, deficiency being common on many acid soils and clays. In addition, inadequate potassium supply is a common occurrence. Deficiencies of the other essential plant nutrients are relatively infrequent or uncommon.

The concentrations of the major nutrients nitrogen and phosphate in substrates which occur on wasteland are usually far below those required for normal plant growth. In contrast, some of the minor nutrients tend to be present in excessive concentrations, iron, manganese, copper, boron and zinc being at toxic levels in some wastes (see Chapter 4).

2.2 Effects of major nutrients on plant growth

(*a*) *Nitrogen* Plants may take up nitrogen from the soil as either ammonium or nitrate ions. Within the plant, nitrogen is converted into aminoacids and thence proteins which results in increased leaf growth and photosynthesis. Plants suffering from a shortage of nitrogen respond to an elevated supply by exhibiting increased leaf greenness, developing larger cells and leaves which have a higher water content. They also show a general improvement in shoot growth with an associated increase in their shoot/root ratios. Tiller production is promoted in grasses.

(*b*) *Phosphorus* Phosphorus is absorbed by plants as inorganic phosphate ions which become involved in enzymic reactions associated with phosphorylation. This nutrient is essential for cell division and plays a crucial role in the development of meristematic tissue.

Phosphate deficiency results in severe stunting of the root systems of plants and top growth is inhibited. The leaves often exhibit a distinctive reddish coloration. The growth of clovers and other legumes is particularly sensitive to shortage of phosphate.

Unlike nitrogen, phosphate increases leaf area without restricting the transport of carbohydrates to the roots. This means that growth is best supported by a balance between nitrogen and phosphate supply. Too much nitrogen will stimulate top growth without promoting a corresponding increase in root development.

(*c*) *Potassium* Potassium plays a part in the synthesis of amino acids and proteins from absorbed ammonium ions and is important in photosynthesis. Like phosphate, inadequate potassium supplies combined with high soil nitrogen status may have detrimental effects on growth and restrict the plant's response to nitrogen. Clovers and allied nitrogen-fixing plants require relatively high amounts of both potassium and phosphate.

2.3 Effects of nutrient additions on plant growth on wastes

The growth responses to additions of major nutrients follow a common pattern on the vast majority of wastes. For example, on colliery spoil (Fig. 2–1) it is clear that applications of either nitrogen or phosphate

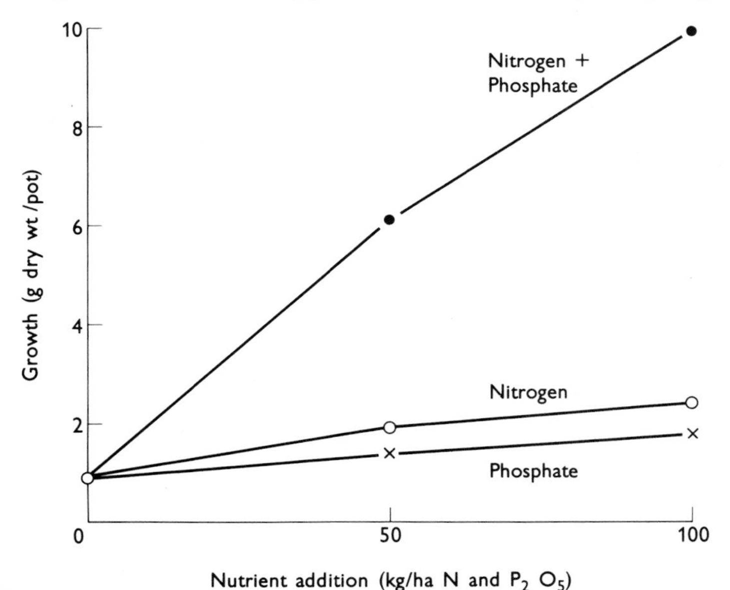

Fig. 2–1 Effects of the nutrients nitrogen and phosphate on growth of ryegrass in limed colliery spoil.

alone produce very poor growth because a shortage of one nutrient prevents a response to the other. This contrasts with the pattern of response on most soils where there is usually sufficient phosphate to sustain very large responses to nitrogen. Thus, on wastes, nitrogen and phosphate interact very strongly to produce massive growth increases when they are supplied in combination (Fig. 2–1).

However, the addition of major nutrients to wastes does not promote growth if toxicity factors such as adverse pH conditions, salinity, or high concentrations of toxic metal ions are present. Under such conditions, major nutrient deficiency is of secondary importance and only has a significant effect when the toxicity factors have been removed. In the absence of specific toxicities, nitrogen and phosphorus deficiencies are nearly always growth-limiting and may be present at concentrations of only 1 ppm as in blast furnace slag (STREET and GOODMAN, 1967).

2.4 Loss of nutrients in wastes

Owing to the inorganic and free-draining nature of many wastes, nutrient retention is often extremely poor and results in the early reappearance of deficiencies. At the same time fixation may occur, resulting in the transformation of nutrients to forms which are unavailable for plant root uptake.

2.4.1 *Leaching of nitrogen and potassium*

Loss of nitrogen by leaching is greatest in wastes which consist of coarse particles and possess high infiltration capacity combined with good internal drainage characteristics. Examples are blast furnace slag and metalliferous smelter wastes (GEMMELL, 1974).

Nitrate nitrogen is more readily leached than ammonium nitrogen because ammonium ions are held strongly by cation exchange capacity. Nitrate ions may be held by anion exchange capacity, but only weakly. Being a cation, potassium is not subject to rapid leaching unless cation exchange capacity is very low or the development of acidity causes a rapid release from clay mineral breakdown as in colliery spoil.

2.4.2 *Volatilization of ammonium nitrogen*

Some chemical wastes contain strong bases such as calcium hydroxide which react with ammonium salts to liberate nitrogen as gaseous ammonia.

$$(NH_4)_2SO_4 + Ca(OH)_2 \rightarrow CaSO_4 + 2H_2O + 2NH_3\uparrow$$

2.4.3 *Fixation of phosphate*

There are four important mechanisms which may prevent plants extracting phosphates from wastes. The first three listed below are a consequence of pyrite oxidation and therefore confined mainly to colliery wastes.

1. Phosphate ions combine with iron in solution to form a highly insoluble precipitate of ferric phosphate (DOUBLEDAY, 1971).
2. Phosphates react with labile aluminium released from clay minerals under low pH conditions to form the highly insoluble aluminium phosphate. In alkaline wastes, labile aluminium may fix phosphate by the same mechanism.
3. Phosphate is adsorbed in large quantities by amorphous ferric hydroxide. In colliery wastes, the ferric hydroxide often forms as a coating over the mineral particles and adsorption may occur under alkaline as well as acid conditions (DOUBLEDAY, 1971).
4. Under alkaline conditions, as in blast furnace slag, phosphate may be immobilized as highly insoluble forms of calcium phosphates (GEMMELL, 1975).

It is believed that the third mechanism is of principal importance in colliery wastes.

Figure 2–2 illustrates how phosphate fixation is related to the iron

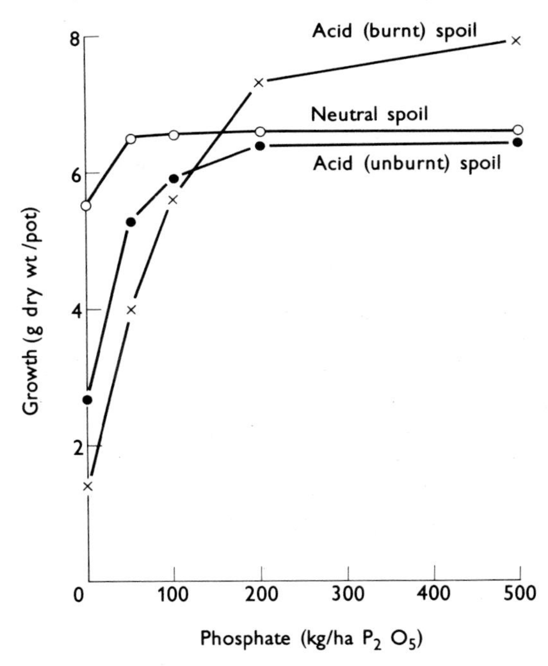

Fig. 2–2 Growth responses to applied phosphate on spoils of differing phosphate fixation capacity.

pyrite content of colliery spoil. The neutral spoil was non-pyritic and did not exhibit phosphate fixation. The acid unburnt spoil was highly pyritic and, because of fixation, required much more phosphate for correction of the deficiency. When this type of spoil is burnt, however, the phosphate requirement is increased because of the fixing power of ferric compounds formed by the complete oxidation of the pyrite. Therefore, in unburnt material, the unoxidized pyrite is not only a source of potential acidity but gives rise to potential phosphate fixation as well.

2.5 Correction of nutrient deficiencies

2.5.1 Nitrogen

Nitrogen deficiency is easily rectified by the application of chemical fertilizers, such as ammonium nitrate or urea, or by spreading organic materials and manures such as farmyard manure, broiler manure, or sewage sludge. Another method is to plant leguminous species which have the power to fix atmospheric nitrogen by means of root nodule bacteria.

As explained earlier, ammonium nitrogen fertilizers are more efficient than nitrates when used on wastes unless the substrate being treated is alkaline. However, when cation exchange capacity is low, even ammonium nitrogen is subject to rapid loss by leaching and may have largely disappeared before seedlings are established. Under such conditions it has been found that recently developed slow-release fertilizers such as sulphur-coated urea are a more efficient nitrogen source. Experiments on blast furnace slag in Cumbria (Table 5) have demonstrated that 50 kg/ha N applied as sulphur-coated urea gives better growth than 100 kg/ha N added as uncoated urea. Nitrogen is released from sulphur-coated urea over a period of about 150 days whereas uncoated urea is highly soluble in water.

Table 5 Effects of sulphur-coated and uncoated ureas on performance of *Festuca rubra* swards on blast furnace slag. (Adapted from GEMMELL, 1974, *Nature*, 247, 199–200.)

Nitrogen fertilizer	*Rate of application kg/ha N*	*Index of visual acceptability (1–10)*
Sulphur-coated urea	100	5·37
Sulphur-coated urea	50	4·73
Uncoated urea	100	4·43
Uncoated urea	50	3·83
Control	Nil	3·30

Least significant difference at P 0.05 = 0.41.

Organic materials are particularly good slow-acting sources of soil nitrogen and for this reason give better and longer-lasting growth responses on wastes than do chemical fertilizers (Table 6). They also improve nutrient retention. If applied together, however, organic and chemical fertilizers have an impressive effect on plant growth (see fertilizer plus sewage sludge in Table 6). It is also clear from the table that nitrogen fixation by clover is a good substitute for chemical and organic fertilizers although it may be necessary to introduce *Rhizobium* bacteria into the wastes before fixation can occur.

Table 6 Growth responses of established grass swards to various sources of soil nitrogen on blast furnace slag.

	Dry weight (g/m²)			
Presence of clover	*Control Nil* N	*Fertilizer 120 kg/ha* N	*Sewage sludge 1.5 cm*	*Fertilizer and sewage sludge*
Grass only	1	80	260	430
Grass/clover	270	290	390	470

The levels of nitrogen fertilizer application used for seeding are normally in the range 30–120 kg/ha N, the rate being adjusted in relation to the amount of growth required, the species planted, and the nature of the substrate. Low rates are used for autumn sowing and when clovers are included in the seeds mixtures. The species most commonly sown for nitrogen fixation are *Trifolium pratense* (red clover) and *Trifolium repens* (white clover); the most persistent clovers in pastures are the wild white varieties of *Trifolium repens*. Some woody plants such as *Alnus spp.* (alders) and *Robinia* (acacia) can fix nitrogen and are good pioneer species for woodland plantings.

2.5.2 *Phosphate*

Unlike nitrogen, phosphates are generally immobile in soils and wastes. Consequently, they are unlikely to be transported by leaching beyond the level of root penetration.

Phosphate deficiency can be rectified by the application of calcium phosphates or by basic slag from the steel-making industry. Rock phosphate is unsuitable because of its very low solubility. Some pyritic colliery shales have been reported to require up to 1000 kg/ha P_2O_5 because of phosphate fixation but the usual requirements for colliery wastes are in the range 200–300 kg/ha P_2O_5. Other types of waste need at least 100 kg/ha P_2O_5.

Root development of grasses and trees may be restricted if phosphates are not incorporated deeply into acid colliery waste substrates. The addition and integration of phosphate is best carried out after liming because it has been found that fixation occurs to a lesser extent after the pH has been raised (DOUBLEDAY, 1972).

2.5.3 *Potassium and other nutrients*

Potassium requirements are generally no greater than 50 kg/ha K_2O. This nutrient is mostly applied in the form of potassium sulphate.

The only other nutrients which are occasionally in short supply in wastes are calcium and magnesium which can be added as calcium carbonate (limestone) and magnesium carbonate (magnesian or dolomitic limestone) respectively. Calcium sulphate may be used to supply calcium if a pH adjustment is unnecessary.

2.6 Maintenance of soil fertility

The establishment of vegetation on wastes can rarely be achieved by a 'once and for all' treatment unless deep soil is provided, usually at great expense. Although vegetation may not thrive owing to the reappearance of toxicities, the most common after-management problem is shortage of nutrients.

Wastes differ from most surface soils in being devoid of organic matter and lacking a proper soil structure. This means that growth is dependent on fertilizer applications until the formation of a reasonable level of humus which then acts as a reservoir of nutrients to replenish losses in the soil solution. Figure 2–3 shows how growth may decline if no further nutrients are supplied but immediately recovers and continues to improve when nutrients are added to restore soil fertility.

Nitrogen is the most rapidly exhausted nutrient because it is prone to leaching. In the case of phosphate, fixation may necessitate more applications of this nutrient during the after-management period.

In general, organic materials are better sources of nitrogen than inorganic nutrients because they release nitrogen slowly, provide humus, and improve the structure of the waste substrate. The conservation of grass/legume associations is of particular importance on reclaimed land as an economical means of maintaining nitrogen fertility.

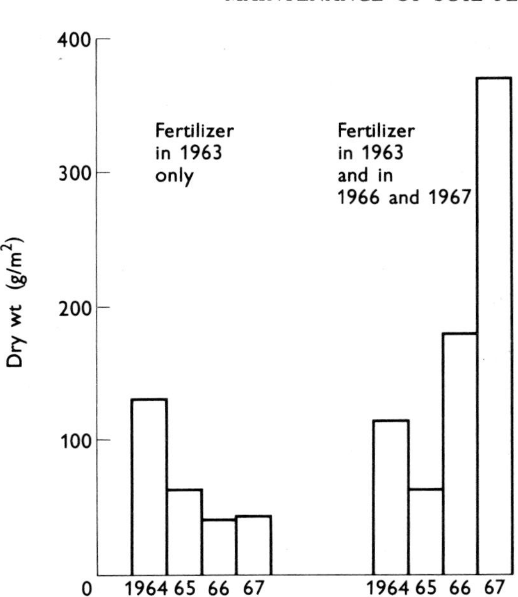

Fig. 2–3 The effect of nutrient applications on the recovery of grass swards on copper smelter wastes. (After GOODMAN *et al.*, 1973; courtesy of Gordon and Breach, New York and London.)

3 pH Conditions and Plant Growth

3.1 Effects of acidity on plant growth

The growth of plants can be affected directly and indirectly by the presence of soil acidity. It appears that if the soil pH is above about 4.0 then acidity is not directly toxic to plant roots, the adverse effects on growth being due to induced nutrient deficiencies. Below pH 4.0, except in the cases of acid tolerant species such as *Deschampsia flexuosa* (wavy hair grass) and *Nardus stricta* (mat grass), the concentration of hydrogen ions begins to be directly harmful and there may also be secondary effects such as aluminium and manganese toxicities. The physical structure of the soil may be changed too, as for example in the development of extreme impermeability in colliery shale due to deflocculation of the clay colloids as acidity increases.

(*a*) *Effects on nutrient uptake* The major effects of acidity on the nutrition of plants are induced shortages of available calcium, phosphate, and in some cases symptoms of molybdenum deficiency. In colliery spoil, high concentrations of iron and aluminium salts become soluble at low pH and cause fixation of phosphate, as described in the previous chapter. The addition of extra phosphate does not solve the deficiency problem until the fixation capacity of the waste has been exhausted and the pH raised by neutralization with lime.

(*b*) *Toxicity caused by acidity* Highly acid conditions bring certain metallic cations and toxic salts into solution. Aluminium and manganese are important in colliery shale where the breakdown of clay minerals liberates substantial quantities of aluminium and manganese ions into solution (CHADWICK *et al.*, 1969). At the same time, the concentrations of other potentially toxic metal cations such as zinc, copper, lead, iron and magnesium are increased.

It is well known that enzyme systems in plants are only operative under certain pH conditions. Hydrogen ion concentrations giving pH values around and below about 3.5 cause inactivation of most enzyme systems, bringing about inhibition of root respiration and restricting the uptake of mineral salts and water. These physiological effects of acidity apply also to micro-organisms so that the decomposition of leaf and root litter is restricted and the acitivty of bacteria normally living in close association with plant roots is inhibited.

3.2 Effects of alkalinity on plant growth

The adverse effects of alkalinity in soils are, like those of acidity, usually due to secondary factors. High alkalinity is rarely a characteristic of natural soils. In moist calcareous soils, for example, the pH does not exceed about 8.4 when the soil is in contact with air containing the same carbon dioxide content as the atmosphere. Only soils with sufficient exchangeable sodium for sodium carbonate to be formed in the soil solution are of higher pH status than this. However, certain industrial wastes contain hydroxyl ions of the strong base calcium hydroxide, a situation which may bring about very severe alkalinity. pH values up to 12.6 have been found in chromate smelter waste and alkali wastes (GEMMELL, 1973).

(*a*) *Effects of alkalinity on nutrient availability* As described in Chapter 2, phosphate is immobilized in alkaline soils through the formation of insoluble calcium phosphates. Other examples of nutrient deficiencies induced by alkalinity are those of iron, manganese, boron, magnesium and possibly other trace elements. Iron deficiency chlorosis is well known in nature on highly calcareous soils. The same phenomenon may occur on colliery spoil treated with heavy doses of liming materials (DOUBLEDAY, 1972).

(*b*) *Toxicity caused by alkalinity* It is possible that extremely high pH conditions may cause aluminate toxicity. The toxic effect of chromate in chromate smelter waste has been found to be intensified by the presence of calcium hydroxide (GEMMELL, 1973) although this may be largely due to solubility effects.

Enzymes are inactivated directly as a result of severe alkalinity which may, therefore, have far-reaching effects on the metabolism of plants. Strong bases also bring about the degradation of organic matter and free ammonia is liberated from ammonium salts.

3.3 pH conditions in different wastes

3.3.1 Acidity in colliery spoil

Acidity is the principal growth-limiting factor in many types of colliery waste (CHADWICK, 1973). The majority of shale heaps from deep coal-mining in Northumberland and Durham, Yorkshire, and the South Lancashire Coalfield consist of moderately to intensely acid spoil. On the other hand, tips in some areas such as South Wales and the Midlands tend to be less acid and are commonly of neutral or alkaline pH.

Spoil pH is subject to enormous variation, even on a single tip site. DOUBLEDAY (1971) measured the % frequency of pH values in shales from the Northumberland and Durham areas and found that the pH of acid tips usually falls within the range 3.0 to 5.0. However, the pH of some types of shale may be as low as 1.5 to 2.0 whereas, at the other extreme, values as high as 8.7 have been encountered. In nearly all cases of intense acidity the shale is rich in iron pyrites.

Numerous factors influence the pH of colliery shale, the most important of which are:

1. The duration of exposure of the waste.
2. The presence and mineral form of iron pyrites and the abundance of aluminium hydroxide.
3. The presence and abundance of acid neutralizing materials such as calcium and magnesium carbonate minerals.
4. Whether combustion has occurred.

The duration of exposure is important because atmospheric moisture, oxygen in the air, and weathering have a profound effect on pyrite oxidation. When the waste is freshly exposed following deep mining or opencast extraction, the pH is usually around neutrality (DOUBLEDAY, 1972). Then, as soon as the iron pyrite in the waste comes into contact with moisture and oxygen, oxidation and hydrolysis commence with the ensuing liberation of hydrogen ions into solution. Various reactions are involved but the changes can be represented simply as follows:

$$4FeS_2 + 15O_2 + 14H_2O \rightarrow 4Fe(OH)_3 + 16H^+ + 8SO_4^{2-}$$

Thus, the pH falls rapidly unless the shale contains neutralizing minerals. The length of time required for the pH to reach a fairly steady level depends mainly on the physico-chemico properties of the spoil and may take anything from a few days to many years.

The nature and mineral form of the iron pyrite in the waste seem to be the principal factors governing the rate and degree of acidification. Some forms of pyrite seem to be quite resistant to oxidation whereas other forms are highly reactive. Studies in the United States by CARUCCIO (1973) have revealed that framboidal pyrite, so called because of its strawberry-like appearance when examined under high power microscopy, is the most reactive form and probably the principal source of acidity in all colliery wastes. Its high reactivity is almost certainly a consequence of its small grain size and large surface area exposed for oxidation, although this does not entirely explain its reactive properties.

If pyritic waste contains carbonate minerals, it is possible that their potential neutralizing capacity may exceed the potential acidity which can be generated from pyrite oxidation and hydrolysis. Usually, however, the converse is the case so that the pH is fairly stable for a time until all the carbonate minerals have been used up in acid reactions after which time a rapid fall in pH occurs.

The combustion of colliery shale results in the complete oxidation of pyrite. Therefore, although the waste may have a very low initial pH, commonly in the range 2.7 to 3.5, potential acidity is absent.

At many collieries several different coal seams were mined either simultaneously or at different times with the result that various types of associated strata were tipped as waste. Further, the nature of the associated strata may change along the length of one seam. Consequently, great variation in the composition of colliery waste often occurs on a single tip site although there are many cases where the waste does not vary to any considerable degree. The latter usually happens when mining has been confined to one particular seam and where the associated strata were deposited in one paleo-environment.

3.3.2 Acidity of metalliferous and other wastes

Metal mine spoils are not usually highly pyritic and thus pose less serious problems of acidity in comparison with colliery wastes. Copper mine spoil is often slightly to moderately acid with pH values in the range 3.0 to 5.0, the pH of the waste at the extensive opencast but now derelict copper mine at Parys Mountain in Anglesey being about 3.5. Lead mine waste may be acid or calcareous in reaction whereas zinc mine spoil is generally of pH 6.0 to 8.0.

STREET and GOODMAN (1967) found that copper smelter wastes in the Lower Swansea Valley were slightly acid at pH 5.5 and the few other comparable wastes in Britain are very similar. Zinc smelter waste was found to be neutral or alkaline, waste from old extraction processes being of pH 6.5 whereas waste from a modern plant had a pH of 9.0.

Whereas acidity in colliery wastes is often the principal factor inhibiting the growth of plants, this is rarely the situation in the case of metalliferous wastes where metal toxicity is of overriding importance. Nevertheless, acidity is important because it raises the solubilities of metal ions and therefore increases toxicity. In copper mine spoils and smelter wastes, the pH may be the factor determining whether copper ions are at toxic concentrations or not.

Non-metalliferous wastes in which acid substrate conditions prevail include foundry sand, furnace ash and cinders, some wastes from sand and gravel extraction, exposed subsoils and clays, and certain types of china clay waste.

3.3.3 Alkalinity of power station ash

Power station ash is often moderately or intensely alkaline, varying in pH from about 7.0 to 12.0 (TOWNSEND and HODGSON, 1973). The high pH is accounted for by the presence of calcium and hydroxide ions resulting from the hydrolysis of calcium oxide and calcium silicate minerals. Even in high pH types of ash, however, alkalinity is rarely the principal cause of toxicity although it contributes to it.

3.3.4 Alkalinity in chemical wastes and blast furnace slag

The strong base calcium hydroxide is a toxic constituent of alkali waste resulting from the obsolete Leblanc Process used during the last century for manufacturing sodium carbonate. In the early part of the present century, the process was replaced by the Ammonia-Soda or Solvay Process which is also contaminated with calcium hydroxide. The Leblanc Process contained, in addition, calcium sulphide which hydrolyses to calcium hydroxide:

$$2CaS + 2H_2O \rightleftharpoons Ca(OH)_2 + Ca(SH)_2$$

The pH of freshly exposed alkali waste is about 12.6 because the moisture present is saturated with calcium hydroxide. In some cases the waste was reprocessed in order to extract sulphur by the Chance-Claus Process which effectively removed the sulphides.

BREEZE (1973) found that calcium hydroxide is also the source of alkalinity in chromate smelter waste from the lime method of extraction. Here, the pH may also be as high as 12.6 but the presence of calcium chromate is of over-riding importance as a toxic substrate factor (GEMMELL, 1973).

Blast furnace slag is another strongly alkaline waste which, when unweathered, is of pH 10.0 to 10.5 (GEMMELL, 1974). The hydroxide ions present are believed to be formed by the hydrolysis of calcium oxide and basic silicates, as in power station ash.

3.3.5 Alkalinity in oil shale and iron ore waste

Oil shale has been reported to be alkaline at pH 8.0, a situation which is unaffected by leaching. Presumably this is due to the presence of carbonate minerals. Iron ore waste may be more alkaline, as a pH of 9.0 has been reported for one waste.

3.4 Production of acidity in pyritic wastes

3.4.1 Effect of weathering on acid generation

The oxidation and hydrolysis of iron pyrite in colliery shale are the causes of long-term acid generation and thus responsible for the regression of growth and dieback which have occurred on many reclaimed pit heaps.

Figure 3–1 illustrates the differences in behaviour of pyritic and non-pyritic colliery wastes following ground limestone applications at 10 000 kg/ha in a pot experiment. The pH of the non-pyritic shales is relatively stable with time but falls dramatically if the wastes contain pyrite. If vegetation is planted on inadequately limed pyritic shale, acidity will intensify in the rooting substratum and eventually cause yield regression, development of bare patches, and ultimately total failure.

It is easy to see from Fig. 3–1 that the intial pH values of the unlimed

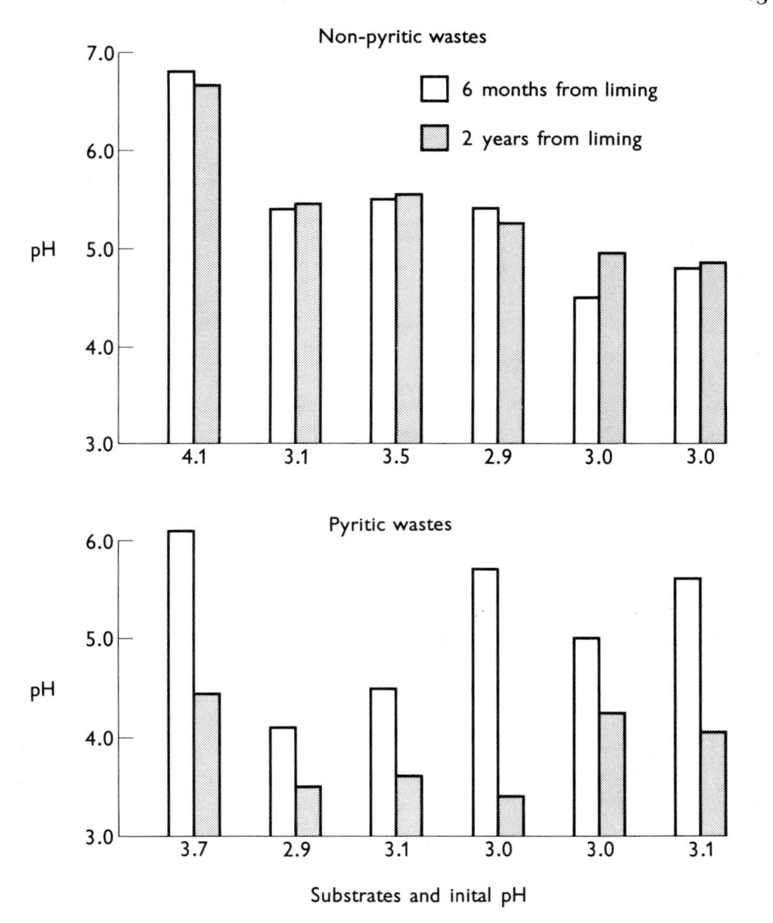

Fig. 3–1 Changes in pH in 6 pyritic and 6 non-pyritic colliery waste samples following limestone application at 10 000 kg/ha.

wastes give no indication of the presence of reactive pyrite. This means that pH determinations and lime-requirement tests cannot be used to estimate the amounts of neutralizing materials needed to establish growth conditions as they would be on normal soils. Still, it has been found in practice that if a shale has a high clay content with a pH below about 3.5, then it is very likely to be pyritic. Otherwise, the ideal solution to the problem is to determine the initial acidity, the amount of pyrite which is capable of oxidation to release acid, and the amounts of carbonate minerals present which are able to neutralize acids generated. In theory this is simple but no reliable method has been discovered for estimating the quantity of pyrite which is sufficiently reactive to undergo natural oxidation and hydrolysis.

The oxidation of pyrite and release of acidity is confined mainly to the surface layers of pit heaps. DOUBLEDAY (1972) has found that the lowest pH values are at 10 to 22 cm depth where the waste is constantly moist and oxygen in the air has access.

3.4.2 The role of bacteria in pyrite oxidation

Studies of the mechanism of pyrite oxidation have revealed that the reactions involved are assisted by bacteria. The organism responsible is *Thiobacillus ferrooxidans* which has the ability to oxidize ferrous iron and sulphur in an acid environment. Although the chemistry of the reactions involved is incompletely understood, it is believed that the following sequence of chemical changes takes place:

1. $2FeS_2 + 7O_2 + 2H_2O \rightarrow 2FeSO_4 + 2H_2SO_4$
2. $4FeSO_4 + O_2 + 2H_2SO_4 \rightarrow 2Fe_2(SO_4)_3 + 2H_2O$
3. $FeS_2 + Fe_2(SO_4)_3 \rightarrow 3FeSO_4 + 2S$
4. $2S + 3O_2 + 2H_2O \rightarrow 2H_2SO_4$

According to LE ROUX (1969) reactions (1) and (3) proceed independently of bacteria but reactions (2) and (4) are unable to operate without their assistance. It has been postulated that the reactions are of a cyclic nature, reaction (3) causing a sulphur-rich layer to be formed on the pyrite particles which slows down the process of pyrite dissolution until the sulphur has been oxidized by reaction (4). The production of ferrous iron by reactions (1) and (3) also retards pyrite oxidation until the iron has been converted to the ferric state by reaction (2).

3·5 pH neutralization: natural and artificial mechanisms

3.5.1 Natural leaching of wastes

Natural leaching of acid pyritic wastes may cause further pollution (Fig. 3–2). Although surface water run-off and seepages from acid coal shale may remove such large quantities of hydrogen ions that drainage waters become grossly polluted (HILL, 1973), there is a rapid replacement of hydrogen ions due to further pyrite oxidation. HALL (1957) has stated that the production of acidity may continue for 100 years and it may take even longer, perhaps hundreds of years, for natural leaching to deplete the reserves of acidity in the surface layers of waste.

In contrast, the high pH conditions prevailing in some wastes are lowered by leaching in a relatively short period of time. Table 7 shows the changes in pH at two blast furnace slag sites after exposure to rainwater percolation. The Askam site consisted of an old tip which had been exposed for tens of years. There, leaching had removed the hydroxyl ions

Fig. 3–2 Acid drainage effluent from a pyritic chemical waste tip. The oxidizing pyrite releases sulphuric acid, ferrous and ferric salts, and toxic heavy metal salts into the effluent. At times, the drainage water has a pH of about 2·0 which has a devastating effect on the natural vegetation surrounding the tip and on stream life. Only acid tolerant grasses and *Juncus* spp. survive in close proximity to the effluent which appears to spread laterally through the soil.

from the surface to a depth of between 15 and 30 cm. The Ulverston waste had been exposed for only six to nine months because of landscape modelling work, leaving the surface contaminated by hydroxyl ions. Within a year, however, leaching had reduced alkalinity sufficiently to permit establishment of plants. On this kind of waste, natural leaching is the only practical method of reducing alkalinity unless the slag is covered with a blanketing non-toxic growth medium such as topsoil or subsoil.

Table 7 pH profiles in blast furnace slag.

Depth (cm)	*Askam site*	*Ulverston site*
0	7.5	8.1
2.5	7.6	9.0
5.0	7.5	10.2
7.5	7.6	10.3
10.0	7.6	10.5
15.0	7.8	10.4
30.0	10.2	10.5

Natural leaching removes the hydroxyl ions in power station ash much more slowly, probably because there are greater reserves of alkalinity in the form of hydrolysing basic silicates. On alkali waste heaps and chromate smelter waste the ameliorating effect of rainfall is slower still because calcium hydroxide and calcium sulphide are present in extremely high concentrations. Table 8 shows that about 40 years of exposure are needed to clear the alkalinity to a depth of about 25 cm and 80 years to leach out the contaminant from the top 60 cm of waste.

Table 8 Variation of pH with depth in alkali waste heaps.

Depth (cm)	*Period of exposure of waste (approx)* 40 *years*	80 *years*
0	7.8	7.4
10	7.7	7.7
20	7.8	7.6
30	8.8	7.5
40	9.5	7.4
50	9.9	7.4
60	12.2	8.2
70	12.1	9.8

3.5.2 *Artificial neutralization of acidity*

The usual method of neutralizing acid soils and peatlands is to apply ground limestone ($CaCO_3$), slaked or hydrated lime ($Ca(OH)_2$), burnt lime (CaO) or dolomitic limestone ($MgCO_3.CaCO_3$) so that the pH is raised to about 6.5. If a rate greater than 10 000 kg/ha is required, it is advisable to apply the neutralizing agent in two or more applications otherwise nutrient deficiencies may be induced.

On acid wastes containing iron pyrite it is often necessary to modify the normal agricultural or horticultural procedures with respect to liming. Because colliery wastes exhibit potential as well as initial acidity, an excess of neutralizing agent is needed in order to allow for acid generation after planting. Under such conditions both hydrated lime and burnt lime are unsuitable because the pH would be raised to too high a level, causing problems associated with alkalinity. Thus ground limestone or dolomitic limestone are employed, although the latter may be unsuitable on some intensely acid wastes because excess magnesium salts may be formed in the soil solution. Magnesium sulphate is highly soluble and may give rise to salinity and hypermagnesia problems (DOUBLEDAY, 1972).

Limestone application rates of between 30 000 and 40 000 kg/ha are not unusual in colliery waste reclamation. Indeed, recent experience indicates that some pyritic shales may require at least 100 000 kg/ha ground limestone for maintaining the pH at a satisfactory level. Unfortunately, however, these massive application rates of limestone induce further planting problems. DOUBLEDAY (1972) has proved in pot experiments that only moderate rates of limestone addition, which bring the seedbed pH above about 6.5, may depress growth on some wastes. He found that the growth responses of grasses to phosphate applications on some site materials were depressed by over 80% after liming to pH 7.8 (Fig. 3–3). No analyses of the site materials were given which could possibly explain why limestone depressed growth on the Shankhouse site

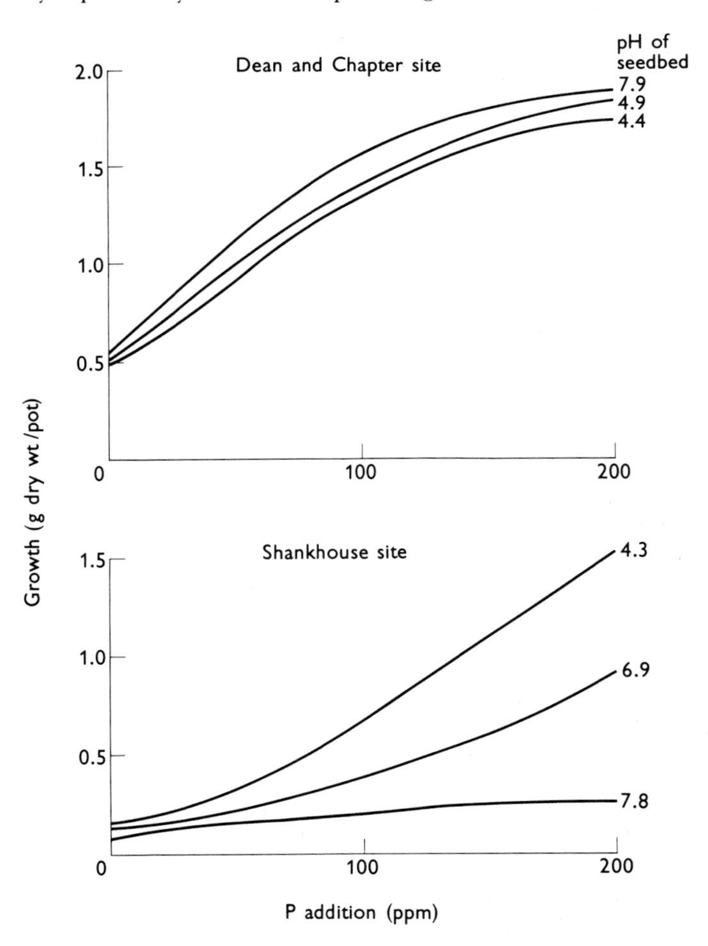

Fig. 3–3 Effect of spoil seed-bed pH on ryegrass growth responses to phosphate on wastes from two sites in North-East England. (From DOUBLEDAY, 1972; courtesy of IPC Business Press Ltd., Guildford, Surrey.)

material but not on the Dean and Chapter spoil. The likely explanation was some kind of trace element deficiency. Evidence in support of this was obtained from further experiments in which Doubleday found that growth on the highly limed Shankhouse waste could be improved by foliar sprays of iron to the leaves, indicating that liming the spoil may induce iron deficiency in the plants. More recent work on Lancashire wastes has shown that clovers are more sensitive than grasses to high liming, the plants exhibiting characteristic symptoms of lime-induced iron deficiency chlorosis.

The problems associated with applying massive rates of limestone cannot be avoided by splitting the applications. If extra limestone is applied as top-dressings after planting, its neutralizing action is confined to the surface layer of the tip soil profile. Consequently, acidity develops unchecked just beneath the surface, with the result that root development is inhibited and ultimately root dieback occurs. Liming materials are immobile in most colliery shales, particularly in those with a high clay content which become extremely impermeable if acidity prevails. Because of this problem, acid colliery spoil is opened-up with rippers prior to planting so that the limestone can be incorporated physically to a given depth and thoroughly integrated with the waste.

A novel method of ameliorating acid waste for colonization is to employ alkaline waste as a substrate neutralizer. Power station ash has been used to ameliorate acid colliery spoil in Germany where lignite-burning power stations produce a highly calcareous ash (KNABE, 1973). Recent investigations in Britain have demonstrated that alkali wastes can be used in a similar way, thus solving two colonization problems at once.

3.5.3 *Artificial leaching of acid wastes*

CRESSWELL (1973) has described how artificial spraying with water has been adopted to leach out acids prior to planting sand dumps from gold mining in South Africa. The success of this method can be attributed to the unusual conditions prevailing in the sands which contain high initial acidity, no potential acidity from pyrite, and are very free-draining.

3.5.4 *Artificial acidification of alkaline wastes*

Sulphur, ferrous sulphate, aluminium sulphate, and dilute sulphuric acid have been employed as soil acidifiers as, for example, in the improvement of alkaline soils of arid regions. Gypsum (calcium sulphate) has been used to treat soils containing excess sodium, the exchangeable sodium being replaced by calcium allowing natural leaching or irrigation treatments to remove the sodium from the surface layers.

The disadvantage of using sulphur is that its oxidation to sulphuric acid is a very slow process in soils, the reaction being dependent on the action of bacteria. The high cost of aluminium sulphate precludes its use when substantial application rates are required. Neither gypsum nor sulphuric acid have been found to be suitable for treating wastes for planting in Britain. Ferrous sulphate and other acidifiers have had little or no effect on toxicity caused by alkalinity in power station ash and cannot be applied to blast furnace slag because they react with calcium carbonate in the waste. However, ferrous sulphate has been employed as a chemical ameliorant during the reclamation of land covered with chromate smelter waste where it was used to treat alkalinity and reduce chemically the toxic levels of chromate present (see Chapter 4). One advantage of ferrous sulphate is that it is an industrial residue from the extraction of ilmenite ($FeTiO_3$) for titanium dioxide and is therefore cheap.

4 Toxicity of Heavy Metals and Dissolved Salts

4.1 Effects of toxic ions on plant growth

(*a*) *Heavy metals and toxic cations* Heavy metals affect plant growth in a number of ways. In solution culture it has been found that exceedingly low concentrations of metal ions, of the order of a few parts per million, cause inhibition of root growth. In soils, their toxicity induces a similar root inhibiting effect, the root systems of afflicted plants being coralloid or stumpy in appearance. Shoot and leaf growth are depressed, the leaves developing acute chlorosis and various types of interveinal necrosis.

The physiological mechanism of heavy metal toxicity is incompletely understood but there is good evidence that most metals produce a very similar kind of metabolic disorder. It has been observed that each metal causes a characteristic type of leaf necrosis or discoloration which is specific to the metal concerned. At the same time, however, a general chlorosis of the younger leaves occurs which is common for all heavy metals. HEWITT (1948) has shown that the latter is an iron-deficiency chlorosis which can be cured by painting the leaves with an iron solution containing a wetting agent. It is thought that the origin of changes occurs in the roots, the plants being unable to utilize the iron content of the roots. The end result of iron-deficiency chlorosis, and therefore of heavy metal toxicity, is a drastic disturbance of cell metabolism, the mineral content of plants being affected in various ways.

The inhibitory effect of heavy metals on root growth seems to be caused by inhibition of cell mitosis in the meristematic zones. Experiments with micro-organisms have also revealed that the cell wall membranes are impaired, allowing potassium and other nutrients to leak out.

The degree of toxicity exerted by different heavy metals varies considerably. Copper is far more toxic than zinc, 0.05 ppm copper being toxic in solution culture whereas the corresponding concentration for zinc is 10 ppm. There are similar differences in the abilities of the metals to produce leaf chlorosis and yield depression. DEKOCK (1956) showed that the order of activity of several heavy metals in producing plant toxicity correlates well with the order of stability of the metallo-organic complexes or chelates which they form. Chelation, in this context, is the holding of toxic metal ions between atoms of complex organic molecules which occur in soil organic matter. It is likely that toxicity is determined by the abilities of the metals to combine with a root protein chelate or similar substance, thus destroying the ability of the chelate to transport iron in the plants. Other types of chelate may also be inactivated.

(*b*) *Salinity* Salinity can affect the growth of plants in two ways, either through raising the osmotic pressure of the soil solution to a high level or by specific effects of the ions concerned.

Elevation of the osmotic pressure of the soil solution to a critical level reduces the power of plants to absorb water. There appears, in general, to be a linear relationship between plant growth and the osmotic pressure of the soil solution. The yields of most crops are reduced considerably when the osmotic pressure of the soil solution exceeds 2 or 3 atmospheres, although salt tolerant species like sugar beet are less affected by such concentrations. Some ions exert specific toxic effects as well. For example, chlorides exert toxic effects on beans at concentrations above 2 atmospheres whereas many grasses are only affected by the osmotic factor. Also, some grasses and certain other plants are affected directly as well as osmotically by sulphate and magnesium ions. Even sodium and calcium may be directly harmful under certain conditions. In addition, most of the anions concerned reduce the availability of phosphate in soils because of solubility effects.

(*c*) *Chromium* Little is known about the behaviour of chromates in soils but their toxic action is undoubtedly connected with their ability to inactivate enzyme systems.

(*d*) *Bicarbonate and lime* Both bicarbonate and excess lime in soils can cause iron deficiency chlorosis as described for heavy metal toxicity. Studies on the bicarbonate ion have revealed that the respiration of root tips is depressed due to inhibition of enzyme systems. Further, both excess bicarbonate and excess calcium carbonate may lead to a deficiency of minor nutrients.

(*e*) *Borates* Information is scant regarding the mechanism of toxicity of borates to plants but it appears to be connected with transpiration. Borates are extremely mobile in plants.

4.2 Heavy metal toxicity in wastes

4.2.1 Metal toxicity in mine spoils

In nature, heavy metals occur mainly as sulphides or carbonates. These compounds are usually insoluble in water but the weathering reactions which occur when metalliferous mine spoils are dumped in the open cause slow oxidation of the sulphides. The oxidation of pyrite as described in Chapter 3 is usually involved, the metals often being present as pyritic minerals such as copper pyrite ($CuFeS_2$). It is now known that the oxidation reactions are microbially assisted and acids are released during weathering causing a marked fall in pH. The result is that the principal metals present in the wastes are released into solution at phytotoxic concentrations.

The total metal content of some typical mine spoil substrates is given in Table 9. It can be seen that the wastes contain high levels of metals other than the ones extracted from the materials. For instance, lead often occurs in association with zinc, and copper mine spoil may contain substantial amounts of zinc and lead. These associated metals may also be present at toxic concentrations.

Table 9 Total metal content of some mine spoils. (After SMITH and BRADSHAW, 1972. *Trans. Inst. Mining and Metall.*, **81**, 230–7.)

	Type of spoil and origin	*Metal content (ppm)* Zn	Pb	Cu	Ca
Zinc mine	(Trelogan, Flintshire)	35 900	10 500	80	74 600
	(Halkyn, Flintshire)	11 700	7400	60	310 500
	(Snailbeach, Shropshire)	20 500	20 900	30	236 900
Copper mine	(Ecton, Staffordshire)	20 200	29 900	15 400	137 100
Lead mine	(Goginan, Cardiganshire)	1400	15 400	210	7
	(Van, Montgomeryshire)	1200	40 500	90	5
	(Glenridding, Westmorland)	2300	5700	40	3800

4.2.2 *Toxicity in smelter wastes*

The extraction of metals from mineral ores by smelting leaves behind considerable amounts of metallic sulphides and other compounds in the slag. Associated metals present in the ores are also released into the slag.

In copper smelter waste, the total residual copper may be present at concentrations in the order of a few per cent. It is normally present as sulphidic compounds in association with ferrous silicates. In zinc slag, a few per cent of total zinc may occur as sulphides in combination with coal and coke ashes and debris from the retort linings. When the wastes are exposed to weathering, the metals are released into solution by microbially assisted oxidation reactions as described for mine spoils. Generally, however, the concentrations of soluble metals are higher than in metal mine spoils with the result that toxicity is of even greater intensity (Fig. 4–1).

Some figures for the levels of soluble metals in the zinc and copper smelter wastes of the Lower Swansea Valley are shown in Table 10. The values for the acetic acid and ethylenediamine-tetraacetic acid (EDTA) soluble metals have to be interpreted with caution because there is now little doubt that these extractants give over-estimates of the levels

Fig. 4–1 Zinc smelter waste tips in the Lower Swansea Valley. Even though some of the tips have been undisturbed for 60 years, zinc toxicity in the waste inhibits colonization by plants except for a few zinc tolerant ecotypes of grasses and micro-organisms.

available to plant roots. The figures given in the table are in the order of hundreds or thousands of parts per million of soluble metals yet concentrations of only a few parts per million are known to be toxic to plant roots in solution culture. Because some grasses are able to grow in the wastes, it is probable that much of the metal concentration dissolved by the soil extractants is not actually plant-available. The water soluble levels, which are also given in Table 10, are far more realistic; water

Table 10 Heavy metal analyses of zinc and copper smelter wastes.

Extractant	*Waste*	Zn	Pb	Cu
Acetic acid (Zn, Pb)	Zinc slag	16 750	4350	850
EDTA (Cu only)	Copper slag	250	15	2042
Water	Zinc slag	140	7	1
	Copper slag	26	0·5	42

extraction techniques are also applicable to metal mine spoils and chemical wastes.

4.2.3 *Toxicity in sewage waste*

Sewage sludge may contain high concentrations of heavy metals derived from industrial effluents and even domestic materials. These metals may be present at such high concentrations that the sludges are toxic to plant growth.

The metals zinc, copper, nickel and sometimes chromium are the major toxic constituents as shown in Table 11. Sludges can usually be

Table 11 Total metal concentrations in some sewage sludges. (After WEBBER, 1972, *Wat. Pollut. Control.*, Metropolitan and Southern Branch, 404–13.)

Type of sludge	*Metal content (ppm)* Zn	Cu	Ni	Cr
Zinc contaminated	48 000	4600	58	60
Copper contaminated	3900	11 300	104	175
Nickel contaminated	5100	670	3900	200
Chromium contaminated	900	795	130	9750
'Uncontaminated'	1000–3000	360–1900	125–300	190–275

classified according to the principal metals which they contain but some, as in the cases of metal mine spoils and smelter wastes, contain two or more metals in phytotoxic concentrations.

4.2.4 Toxicity in slurry from fluorspar extraction

One of the principal factors limiting plant growth on fluorspar tailings is heavy metal toxicity. The mineral fluorspar is frequently contaminated with metallic oxides and small amounts of metallic sulphides. These become concentrated in the slurry and the metals are released as soluble salts by the actions of weathering and oxidation.

4.2.5 Toxicity caused by aerial pollution

Metal pollutants transmitted by air may cause extensive damage to soils and vegetation. In the Lower Swansea Valley, smelter smoke laden with toxic heavy metals from the old smelters has contaminated the natural soils of the valley sides with zinc, copper and lead (STREET and GOODMAN, 1967). Smelting there has now ceased but this kind of pollution still occurs close to some copper refining plants throughout the country.

4·3 Toxicity caused by soluble salts

4.3.1 Chromates in chemical waste

Chromate, present as calcium chromate, is the most important phytotoxin in chromate smelter waste. Apart from chromate, this waste contains various soluble salts (Table 12) which contribute to the toxicity of the material. Investigations on the growth of plants in wastes of different chemical composition have revealed that the intensity of chromate toxicity is influenced by the amount of calcium hydroxide

Table 12 Chemical analysis of chromate smelter waste. (After GEMMELL, 1973, *Environ. Pollut.*, **5**, 181–97.)

Contaminant	*Concentration (ppm)* Unweathered waste	Weathered waste
	Unweathered waste	*Weathered waste*
Chromate	3300 ± 98%	270 ± 81%
Hydroxide	1430 ± 199%	18 ± 364%
Sulphate	25 540 ± 75%	4050 ± 150%
'Bicarbonate'	2440 ± 68%	2745 ± 87%
Calcium	9160 ± 76%	1200 ± 227%
Magnesium	113 ± 304%	1350 ± 56%
Chloride	142 ± 275%	21 ± 33%
pH (1:1 extract)	9·6 ± 1·2	8·6 ± 0·3
Conductance (mmhos/cm of 1:10 extract)	1·4 ± 101%	0·4 ± 91%

Notes: 1. Sodium, potassium and aluminate were present at very low levels.
2. Analyses were conducted on water extracts of the waste.

present, high alkalinity causing greater toxicity than if strong base is low or absent. It is believed that solubility effects play a significant role in determining the overall toxicity when mixtures of different ions are in solution.

Chromates, like most soluble salts, are mobile in soil. It is evident from Table 12 that the concentration of chromate is much lower in the weathered than in the unweathered waste because of leaching. This is also true for sulphate, hydroxide, chloride and calcium ions. As expected, the soluble salt levels as indicated by conductivity determinations were lower at the surface than in the unweathered waste. This movement of soluble salts may actually change the nature of the toxicity at the surface. As natural leaching proceeds, the concentrations of chromate and hydroxide are lowered, but the falling pH causes magnesium hydroxide to become soluble with the result that magnesium ions become toxic in the waste.

A feature of the chemical composition of chromate smelter waste, and indeed of many chemical wastes, is the extreme variation in the amounts of the various ions present. As shown in Table 12, variation may be in the order of several hundred per cent.

4.3.2 Borate in power station ash

The toxicity of power station ash was at one time attributed to excesses of aluminium and manganese salts because plants grown in the ash took up high levels of aluminium and manganese into their foliage and exhibited toxicity symptoms similar to those caused by these two elements (REES and SIDRAK, 1956). However, research at Leeds University by HOLLIDAY, *et al.* (1958) conflicted with the Birmingham studies in that firm evidence was obtained that the causal factor was excess borate. Records of crop tolerances to ash and boron were strikingly similar, as were leaf symptoms of toxicity. Direct agronomic tests and chemical analyses of ash and plants confirmed borate toxicity. As in the case of heavy metal wastes, water soluble determinations of the toxic ion were employed during ash analysis, the concentrations of boron obtained being mainly in the range 20 to 50 ppm (Table 13).

Table 13 Levels of salinity and soluble boron which cause toxicity in power station ash. (After HODGSON and TOWNSEND 1973, *Ecology and Reclamation of Devastated Land*, Vol. 2, Gordon and Breach, New York and London.)

Salinity (mmhos)	*Degree of toxicity*	*Boron (ppm)*
Below 3	Non-toxic	Below 4
4–6	Slightly toxic	4–10
6–10	Moderately toxic	10–30
Over 10	Highly toxic	Over 30

4.3.3 Salinity in power station ash and colliery wastes

Table 13 also gives salinity data for power station ash. The soluble salts concerned are mainly those of sodium and potassium. Although they contribute to toxicity through osmotic effects, they are usually secondary to boron toxicity. Nevertheless, boron concentrations, salinity and pH vary independently in different ash types so that the relative importances of the three factors on different ash sites are subject to great variation.

DOUBLEDAY (1971) has reported that salinity is encountered in some colliery shales and occasionally may take the place of acidity as the principal factor restricting plant growth. Most of the salts are of marine origin and include the chlorides of sodium, potassium and magnesium, as well as calcium and magnesium sulphates. Salinity may be a severe problem in oil shales too (SCHMEHL and McCASLIN, 1973) but little research has been carried out on this kind of waste.

4·4 Amelioration of toxicity caused by ions in solution

4.4.1 Natural leaching

Natural leaching is extremely effective in removing toxicity caused by non-specific salt or osmotic effects. Saline colliery shale can be improved simply by ripping and allowing a period of exposure through the winter months (DOUBLEDAY, 1971) and the results of trials on oil shale in the United States at Colorado have demonstrated the need for natural leaching before colonization can occur As illustrated in Fig. 4–2, NPK fertilization of the shale before leaching did not permit satisfactory

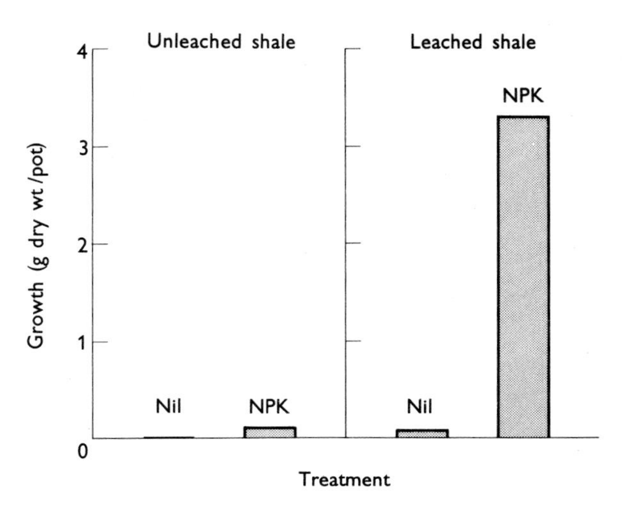

Fig. 4–2 Responses of wheatgrass to NPK fertilizer on leached and unleached oil shale. (After SCHMEHL and MCCASLIN, 1973; courtesy of Gordon and Breach, New York and London.)

growth because salinity was growth limiting. After natural leaching, the majority of soluble salts had disappeared allowing a massive growth response to NPK fertilizer.

The influence of salinity on the toxicity of power station ash is also alleviated by natural leaching. However, because there are considerable reserves of alkalinity and boron, resulting from the hydrolysis of calcium silicate and boro-silicate minerals respectively, the overall toxicity decreases very slowly.

An unusual way in which excess boron can be removed from power station ash is by 'cropping-off' (HODGSON and TOWNSEND, 1973). Clovers in particular take up high levels of boron into their foliage. If the herbage is cut and destroyed rather than ploughed in, ash toxicity is reduced and subsequent growth significantly improved.

Another material which sometimes contains high boron is municipal waste containing cardboard and paper materials. Borates in this waste are readily removed by leaching.

Although natural leaching brings about a dramatic reduction in the surface concentration of chromate in chromate smelter waste (Table 12), toxicity persists because further chromate ions and magnesium are released into solution. The situation with regard to toxic heavy metals in mine spoils and smelter wastes is similar; hundreds or even thousands of years may pass before the metal concentrations become low enough for colonization to occur.

4.4.2 *Effect of organic matter*

As mentioned at the beginning of this chapter, organic matter has the ability to form stable chelate complexes with toxic metal cations, thereby reducing their availability and toxicity to plants in soil. The toxicity of metalliferous wastes can be combated by incorporating materials which possess a high organic matter content (STREET and GOODMAN, 1967); peat, sewage sludge and domestic refuse are good examples of such materials. If sewage sludge is used, it should be relatively low in toxic metal content and be from the 'activated sludge' process. Figure 4–3 shows the effects on growth following the addition and incorporation of 5 cm layers of sewage

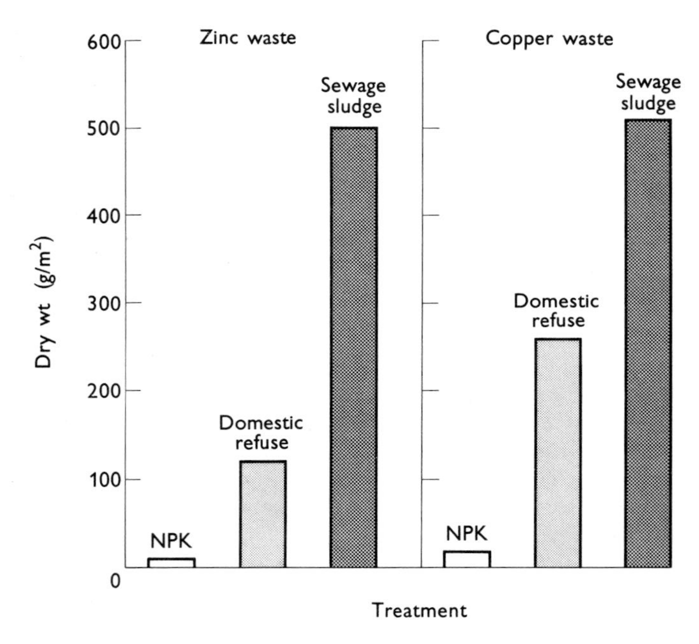

Fig. 4–3 Effects of amendment materials on the growth of grass on smelter wastes in the Lower Swansea Valley.

sludge and domestic refuse into zinc and copper smelter wastes in the Lower Swansea Valley. In this case, sewage sludge was rather more effective than domestic refuse because it possessed a higher organic matter content. Both materials were also rich in major nutrients and produced growth on the wastes comparable with that of normal pastures. The growth effects of the two amendments were a reflection of the soluble levels of metals present in the wastes after amelioration (Table 14).

Organic matter, particularly in the form of sewage sludge, is a highly

Table 14 Reduction in acetic acid soluble metal levels in smelter wastes by organic amendment materials.

		Metal concentration (ppm)		
Type of waste	*Metal*	*No amendment*	*Domestic refuse*	*Sewage sludge*
Zinc waste	Zn	16 750	12 370	6590
	Pb	4350	3170	780
Copper waste	Cu	2040	510	310

successful amendment for use on power station ash (HODGSON and TOWNSEND, 1973). Here, its success can be attributed to a lowering of boron and soluble salt levels by ion association processes and possibly chelation.

4.4.3 *Chemical reduction of chromates*

Toxicity in chromate smelter waste may be alleviated by chemical reduction of the soluble chromate present to chromic salts, thereby lowering the solubility and toxicity of chromium in the waste. Chromic hydroxide is precipitated, chromium in this form being relatively harmless to plants provided that the pH conditions in the substrate are around neutrality.

Ferrous sulphate is an excellent chemical reducing agent for use in this situation and has the additional effect of lowering the high pH of the waste (GEMMELL, 1972). It is available at low cost for it is a residue or by-product from the manufacture of titanium dioxide as mentioned in Chapter 3. The chemical changes which occur when it is applied to chromate smelter waste follow the well known reaction:

$$CaCr_2O_7 + 7H_2SO_4 + 6FeSO_4 \rightarrow Cr_2(SO_4)_3 + 3Fe_2(SO_4)_3 + CaSO_4 + 7H_2O$$

The method is most effective when used in conjunction with organic matter amelioration as shown in Fig. 4–4. It can be seen that peat is far more effective than soil because of its higher organic matter content. The mechanism of action of the organic matter is incompletely understood but it seems that it contributes to the chemical reducing process and lowers the solubility of chromate by ion association effects. It has been confirmed by chemical analysis that the addition of organic matter lowers the concentration of soluble chromate (GEMMELL, 1974) and probably removes other anions and cations from the soil solution by complex formation and ion association mechanisms, thereby alleviating secondary toxicities.

4.4.4 *Ion antagonisms and pH effects*

WILKINS (1957) found that the toxicity of many heavy metal cations in

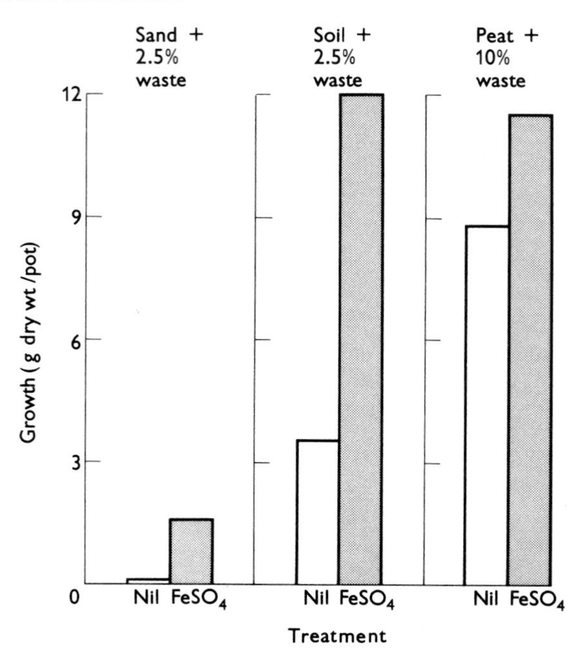

Fig. 4–4 Effect of ferrous sulphate on toxicity of media contaminated with chromate smelter waste. (After GEMMELL, 1972, *Nature*, **240**, 569–71.)

solution is alleviated by the presence of calcium ions as calcium nitrate. This is known as ion antagonism (Fig. 4–5) and explains the low intensity of toxicity of many metals present in calcium rich wastes (see Table 9). Chromate toxicity is also reduced by the presence of calcium ions. The effect in most cases appears to be independent of pH changes and solubility phenomena. It is sometimes possible to apply calcium carbonate or calcium sulphate to wastes as toxicity ameliorants, but their effects are often of small magnitude unless the wastes treated contain little or no initial calcium.

When low pH conditions are present in metalliferous wastes, it is sometimes possible to alleviate metal toxicity by raising the pH by means of lime application. As described in Chapter 3, acidity causes the release of abundant metallic cations into solution from the exchange complex as a result of displacement by hydrogen ions.

4·5 Physical isolation

Techniques based on physical isolation may be used to eliminate toxicity in most situations. In practice, however, such methods are extremely

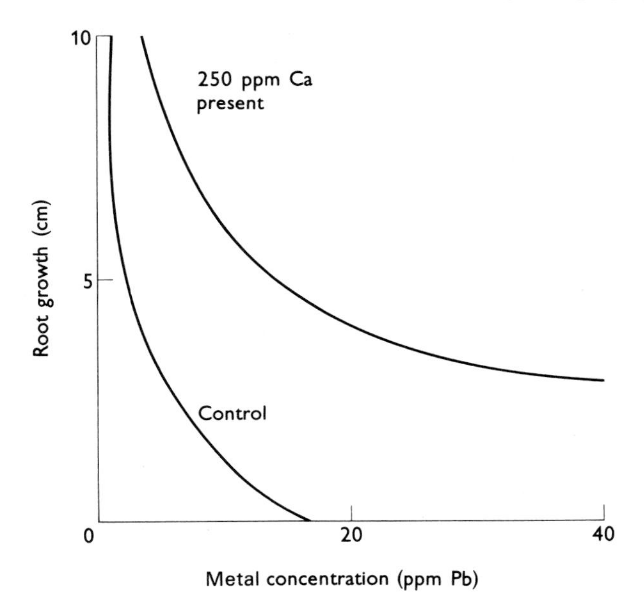

Fig. 4–5 Antagonistic effect of calcium ions on lead toxicity to root growth in solution culture.

expensive and only useful if other amelioration techniques are impractical.

One situation in which physical isolation is a necessity is when highly toxic types of power station ash are reclaimed for agriculture. HODGSON, HOLLIDAY and COPE (1963) demonstrated that a 30 cm depth of soil is required for high yields of arable crops or high intensity grazing. From Fig. 4–6 it can be seen that chemical fertilizers applied at high rates to unsoiled ash will raise yields to only one third of those on soil. If a 30 cm depth of soil is used, normal fertilization rates will achieve high yields. For intensely toxic forms of ash it may be necessary to cover the waste with a 30 cm thickness of subsoil before spreading the final covering of topsoil. If, on the other hand, the ash is reclaimed for amenity, recreation, or low intensity grazing, 5 to 10 cm of soil cover is sufficient provided that high rates of nitrogen and phosphate fertilizers are applied. If topsoil is unavailable or in short supply, subsoil, peat or neutral colliery shale are acceptable substitutes (Fig. 4–7). Sewage sludge and acid colliery spoil can give even better results than soil.

Trials with different cover materials have also been carried out on metalliferous smelter wastes in the Lower Swansea Valley. There, power station ash of low boron content was spread at thicknesses up to 25 cm in order to provide a rooting medium for growth so that the plants did not

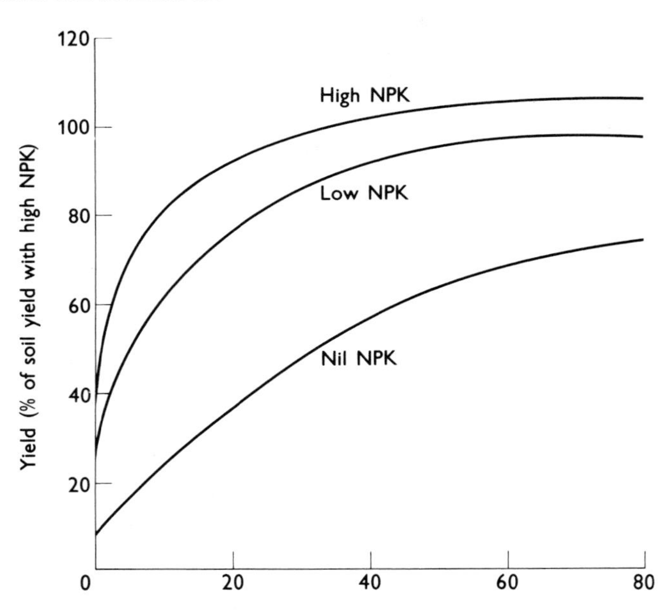

Fig. 4–6 Soil response curves for growth on amended power station ash at three NPK fertilizer regimes. (After HODGSON, 1961, Ph.D thesis, University of Leeds.)

come into contact with the underlying toxic metal waste. GOODMAN, PITCAIRN and GEMMELL (1973) found that layers at least 7.5 cm deep were required for satisfactory growth over the zinc and copper smelter wastes.

A good case can be made for soiling highly acid types of colliery spoil which continue to generate acidity because of pyrite oxidation. The information gained from research on power station ash is particularly relevant for soiling colliery spoil and other types of waste.

The physical isolation of toxic materials like chemical wastes may be absolutely essential for colonization by plants. For example, the toxicity of chromate smelter waste is so severe that massive application rates of ferrous sulphate and sewage sludge are required for satisfactory amelioration. It is more economical to spread an initial covering of subsoil about 25 cm deep for grass and 200 cm deep for trees in order to isolate most of the waste, and to follow this up by applying ferrous sulphate and sewage sludge in order to counteract any residual toxicity caused by contamination of the cover material by the underlying waste. Physical isolation is also necessary because upward mobility of chromate anions may occur, leading to contamination of the rooting zone (BREEZE, 1973). Thus it is essential to amend the chemical waste with a porous,

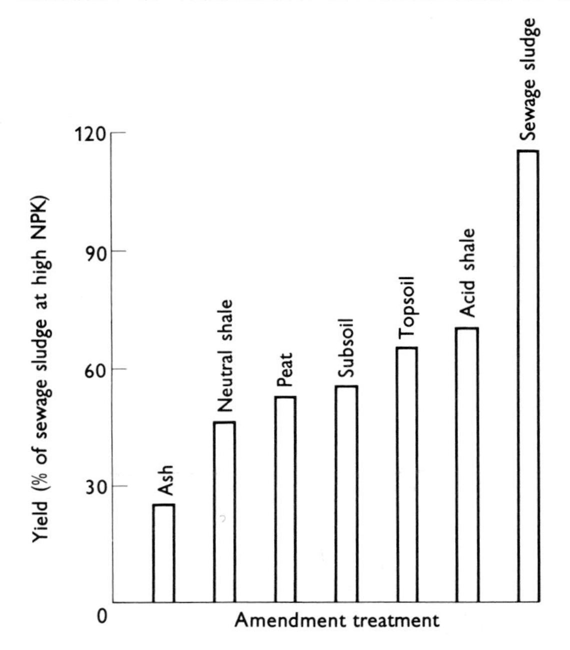

Fig. 4–7 Effect of amendments applied at 7.5 cm depth on growth of grass on power station ash amended with NPK fertilizer. (After HODGSON, 1961, Ph.D thesis, University of Leeds.)

stony material, free of organic matter and clay, so that upward translocation of chromate anions on the soil exchange complex is eliminated. Details of the method used by Lancashire County Council to establish grassland and trees on chromate smelter waste near Bolton are summarized in Table 15.

4.6 Stability of vegetation on treated metalliferous wastes

The long-term production of acidity caused by pyrite oxidation in colliery wastes has been described in Chapter 3. In metal mine spoils and smelter wastes a comparable process occurs in which the metallic sulphides are oxidized to release toxic metal cations into solution. In the presence of organic matter, these cations are immediately complexed and rendered unavailable to plant roots. Eventually, however, the complexing or chelating power of the organic matter may be saturated so that the concentration of metal ions in the soil solution increases to a toxic level. Oxidation and disappearance of the organic matter will, of course, bring about a similar and more rapid increase in toxicity.

Table 15 Summary of treatment method for establishment of vegetation on chromate smelter waste

Amendments (in order of addition)	*Rate*	*Method*	*Effect*
Porous subsoil	25–30 cm depth for grass 200 cm for trees	Spread on surface of waste after completion of levelling and grading	Provides rooting medium and restricts mobility of chromate
Ferrous sulphate ($FeSO_4 7H_2O$)	up to 40 t/ha	Spread on surface and incorporated by discing and harrowing	Chemical reduction of chromate and neutralization of alkalinity
Period of exposure allowed for natural leaching of ferrous sulphate into substratum prior to organic matter amelioration and planting			
Organic matter (peat or sewage sludge)	100 t/ha or more	Spread on surface and incorporated by discing and harrowing	Lowers concentrations of chromate and other toxic salts. Provides nutrients and humus

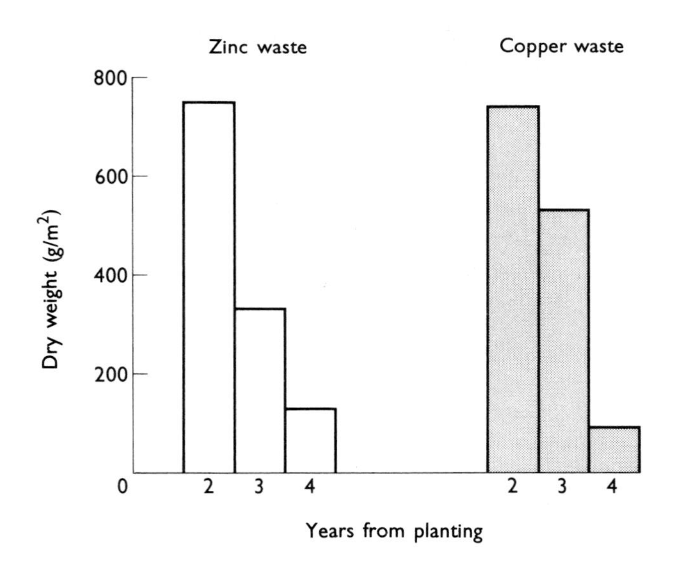

Fig. 4–8 Recession of grass swards due to the reappearance of metal toxicities on organically amended metalliferous smelter wastes.

GOODMAN, PITCAIRN and GEMMELL (1973) found that as a consequence of these changes vegetation established on such wastes begins to deteriorate (Fig. 4–8) and may ultimately fail completely. The only way in which regression may be arrested is by adding further organic materials or by planting metal tolerant ecotypes of grasses as will be described in Chapter 5.

5 Natural Colonization and Plant Tolerance

BRADSHAW (1970) has drawn attention to the fact that man-made habitats can be extremely interesting and valuable scientifically, particularly in studies of evolutionary processes, for natural selection plays an important role in the colonization of extreme habitats. He considers that artificial environments like old mine sites can give great rewards to researchers in the fields of genetics and ecology because the habitats are distinctive, definable, often of a known age, and ecologically simple. In addition, many of the factors such as competition, biotic effects and prehistory which are so complicated in natural plant communities, do not apply on mine wastes.

More recently, KELCEY (1975) has shown how industrial development has provided special habitats (Table 16) which can make a positive contribution to wildlife conservation. He points out that industrial processes have brought about habitat diversification and cites, as an example, the limebed complexes in Cheshire where NEWTON (1971) recorded 24 species of flowering plants which are otherwise rare or unusual in the county.

5.1 Interspecific tolerance

In nature, contrasting habitats such as acid heathlands, sand dunes, chalk downs, fens and marshes are colonized by their own characteristic associations of plant species. The species occurring on a particular habitat are adapted to the edaphic or soil conditions prevailing there. For example, the associations of acid heathlands consist of species tolerant of low pH conditions such as *Calluna vulgaris* (ling) and *Deschampsia flexuosa* (wavy hair grass). *Ammophila arenaria* (marram grass) is an example of a species possessing a highly specialized kind of tolerance, being adapted to the shifting and accumulating substrates of unstable sand dunes. Likewise, the species which colonize industrial wasteland are limited to those which can tolerate the adverse physico-chemical growing conditions of the wastes concerned.

5.1.1 Tolerance of low nutrient conditions

In the absence of toxic substrate factors, the flora of industrial wasteland is influenced primarily by nutrient conditions. As described in Chapter 2, the habitats are predominantly nutrient-poor with low nitrogen and phosphate regimes. Thus there is an intense selection pressure for species with low nitrogen and phosphate demands like

Table 16 Classification of industrial habitats. (From KELCEY, 1975) *Environ. Conserv.*, **2**, 99–108.)

1. *Discrete sites:*			
(a) Extractive	(i)	Clay pits	wet
			dry
			spoil heaps
	(ii)	Gravel pits	wet
			dry
	(iii)	Sand pits	wet
			dry
	(iv)	Chalk quarries	
	(v)	Other limestone quarries	
	(vi)	Mineral workings	
	(vii)	Quarries in igneous rocks	
	(viii)	Quarries in sedimentary rocks (other than chalk and limestone)	
(b) Industrial *per se*	(i)	Industrial waste	
	(ii)	Flashes (or subsidence lagoons)	
	(iii)	Mill races	
	(iv)	Grounds of industrial complexes	
(c) Miscellaneous	(i)	Sewage works	
	(ii)	Sludge lagoons	
2. *Linear systems:*			
(a) Canals, used and disused, including the verges, bridges, locks and lock gates.			
(b) Railways, used and disused, including the verges, ballast, bridges, walls, platforms, etc.			
(c) Roads	(i)	Ancient trackways	
	(ii)	Classified and unclassified road verges	
	(iii)	Motorways	

Agrostis tenuis (common bent), *Festuca rubra* (red fescue) and *Festuca ovina* (sheeps' fescue). These species are frequently dominant or co-dominant in grass-herb communities of derelict land. The common agricultural grasses such as *Lolium perenne* (perennial ryegrass), *Festuca pratense* (meadow fescue) and *Phleum pratense* (timothy) possess high nutrient demands which accounts for their low incidence on industrial wastelands.

Legume species are adapted to low nitrogen conditions by virtue of their nitrogen-fixing ability but most are depressed by low phosphate and low calcium levels. Exceptions are *Lotus corniculatus* (common birdsfoot trefoil) and *Trifolium dubium* (common yellow trefoil) which can withstand low levels of these nutrients and are therefore well adapted to wasteland habitats.

5.1.2 *Tolerance of extreme pH conditions*

Deschampsia flexuosa has formed a complete grass cover on many acidic colliery waste heaps throughout the country. It will tolerate acid spoils of pH 3.2 and is able to respond to additions of nitrogen and phosphate when growing in unlimed and highly acid pit soils (DAVISON and JEFFERIES, 1966). Wastes of about pH 4.0 and above generally support a number of acid tolerant grasses and herbs but the communities are typically species-poor. In contrast, and as occurs in nature on base-rich soils, alkaline and calcareous wastes support species-rich plant communities. However, extreme basicity restricts the number of species which can survive so that only alkaline tolerant species like *Festuca rubra* are to be found if the pH is above about 8.5.

Many species which appear naturally on industrial wastes are either calcifuges or calcicoles, depending on whether the substrates are acidic or basic respectively. Since calcifuges and calcicoles are, respectively, intolerant of basic and acid conditions, the nature of the flora can give a good indication of the pH status of the wasteland soils. Some useful indicator grass species are listed in Table 17.

Table 17 Principal grass colonizers of acid and alkaline industrial waste habitats.

Species	*Preferred soil reaction*	*Notes*
Deschampsia flexuosa	Acid (calcifuge)	Tolerant of highly acid colliery wastes, copper slag and cinders. Often sole colonizer and dominant
Nardus stricta	Acid (calcifuge)	Locally dominant on acid colliery wastes and cinders
Agrostis tenuis	Acid or neutral (avoids high pH)	Very common and locally dominant on acid wastes but occurs on neutral wastes too
Holcus lanatus	Acid or neutral (stimulated by low pH)	Common on acid colliery spoil but is frequent in a variety of neutral waste habitats
Dactylis glomerata	Alkaline (calcicole)	Often co-dominant with *Festuca rubra* on alkaline wastes but is also frequent on neutral materials
Festuca rubra	Alkaline (not a calcicole but more frequent on alkaline wastes)	Tolerant of high pH. Dominant on alkaline wastes, blast furnace slag and calcareous mine spoils but may occur on neutral and acid wastes
Festuca ovina	Acid or alkaline	Dominant or co-dominant with *Festuca rubra* on alkaline or calcareous wastes

5.1.3 *Tolerance of dissolved salts*

It is well known that many maritime species are adapted to survival in areas with high concentrations of soluble salts. Examples are species of the genera *Atriplex* and *Chenopodium*. The salt tolerance of such species enables them to grow on saline industrial wastes like power station ash.

As mentioned in the previous chapter, HOLLIDAY *et al.* (1958) discovered that boron as borate could be a toxic factor in power station ash. They classified a number of species according to their tolerance of boron in the ash. Of the common grasses and legumes, most grasses are of medium or low tolerance whereas clovers exhibit a considerable range of tolerance (Table 18).

Table 18 Tolerance of some grasses, clovers and trees to power station ash. (Adapted from HODGSON and TOWNSEND, 1973, *Ecology and Reclamation of Devastated Land*, Vol. 2, Gordon and Breach, New York and London.)

Species	*Tolerance*	*Species*	*Tolerance*
Grasses		Legumes (cont.)	
Lolium perenne	ST/T	*Trifolium fragiferum*	S
Lolium multiflorum	ST/T	*Trifolium dubium*	S
Poa spp.	ST/T		
Festuca rubra	ST/T	Trees	
Dactylis glomerata	ST	*Alnus glutinosa*	T
Agrostis spp.	ST	*Populus alba*	T
Phleum pratense	S/ST	*Salix* spp.	T
Cynosurus cristatus	S/ST	*Acer pseudoplatanus*	ST
Festuca pratense	S/ST	*Betula verrucosa*	ST
		Robinia pseudoacacia	ST
Legumes		*Fagus sylvatica*	S
Melilotus alba	T	*Fraxinus excelsior*	S
Trifolium repens	ST/T		
Trifolium pratense	ST/T		

T=tolerant; ST=semi-tolerant; and S=sensitive.

Little information is available concerning the resistance of different species to specific cations or anions. Certain species possess tolerance to metal-rich soils but the mechanism of resistance is unclear. REPP (1973) gives a list of species found on highly metalliferous soils throughout Europe. For example, *Linum catharticum, Thlaspi alpestre* subsp. *calaminaria, Plantago lanceolata* and *Viola lutea* can tolerate up to 6% zinc in the soil. Most of the common species, however, do not appear to differ appreciably in their general ability to tolerate toxic levels of metallic ions in soils, although certain species exhibit significant variations or ecotypic differences as will be described in section 5.2.

5.1.4 *Tolerance of adverse physical and moisture conditions*

Industrial environments pose various physical problems such as steep slopes, sharp drainage conditions, waterlogging with some land below the water table, rocky and stony substrata, compaction of the ground materials and concretions in some wastes. It is generally true to say that nearly all types of industrial environment can be matched in nature. For example, natural rocky and stony habitats are cliffs and screes and many waterlogged industrial environments resemble marshland.

Because of its similarity to natural habitats in many respects, colonization of industrial areas occurs by migration of adapted or tolerant species from the nearest suitable natural habitat communities. Thus alders, willows and sallows which are tolerant of impermeable and waterlogged ground conditions quickly colonize large areas of wasteland where such conditions prevail. Reed-beds develop on mining subsidence areas and in clay pits. Coal mining subsidence flashes, flooded gravel pits and sand pits are colonized by aquatic fauna and flora. Indeed, many of these sites have developed into areas of considerable ornithological and botanical interest and some have been designated as nature reserves.

Nevertheless, some wasteland presents conditions too hostile even for the most adaptable species. Slate quarry waste is a case in point where the physical nature of the material prevents colonization. Concretions caused by cementation processes in power station ash cannot support any species and must be broken down artificially by deep subsoiling before colonization can occur. Natural colonization may be inhibited by chemical toxicities, in which case recourse must be made to the improvement treatments detailed in the preceding chapters.

5.2 Tolerant ecotypes

During studies of *Agrostis tenuis* populations from different habitats, BRADSHAW (1952) discovered that plants collected from an old lead mine site in Wales grew normally on the mine waste whereas plants from uncontaminated sites died. This resistance to the waste was permanent and was presumed to be genetically controlled because it did not change over a two-year observation period when the plants were grown in soil. Since Bradshaw's discovery, other species have been found to possess a similar tolerance (GREGORY and BRADSHAW, 1965) although the phenomenon is restricted to a few out-breeding species like *Agrostis tenuis* which exhibit great natural variation and differentiation into local populations (Table 19).

Table 19 Species of grasses that have ecotypes exhibiting tolerance to metals found in non-ferrous metal mine spoils.

Species	*Tolerance*
Agrostis tenuis	Zn, Cu, Pb, Ni
Agrostis stolonifera	Zn, Cu, Pb, Ni
Agrostis canina	Zn
Festuca rubra	Zn, Pb, Cu
Festuca ovina	Zn, Pb
Anthoxanthum odoratum	Zn
Holcus lanatus	Zn

5.2.1 *Heavy metal tolerance*

The tolerance of plants to heavy metal ions can be measured by a root elongation technique which was devised by WILKINS (1957). He found that only 3 ppm Pb as lead nitrate was toxic to root growth of *Festuca ovina* in solution culture. Because this exceedingly low concentration of the heavy metal cation is very difficult to maintain in solution on account of adsorption effects onto glassware and plant tissues, Wilkins devised a method of working with a higher concentration using calcium nitrate as an antagonistic agent (see section *4.4.4*). The presence of this salt at a concentration of about 1 g/l ($Ca(NO_3)_2 4H_2O$) reduces the toxicity of lead by up to twenty times, so that 25ppm Pb may be used to assay plant tolerance. Calcium nitrate is antagonistic to the toxicities of zinc, copper and other heavy metals.

The technique involves placing single grass tillers in glass tubes (Fig. 5–1) with the lower parts of the tubes immersed in nutrient solution. When the tillers have rooted, the tubes are immersed in calcium nitrate solution and root elongation is measured over a two-day growth period. They are then transferred to a solution of the appropriate heavy metal salt made up in calcium nitrate and root elongation is measured over a further two-day period. In all cases, the growth of a single root is followed. At least twelve replicate tillers should be examined because the results of the test tend to be variable. The degree of tolerance is expressed as an index which is calculated as follows:

$$\text{Index of tolerance} = 1 + \log \frac{A}{B}$$

where A = growth over first 2 days in calcium nitrate,
and B = growth over second 2 days in heavy metal solution.

This method of assaying tolerance was used by JOWETT (1958) to prove that tolerance is specific to certain metals. Tolerance of one metal does

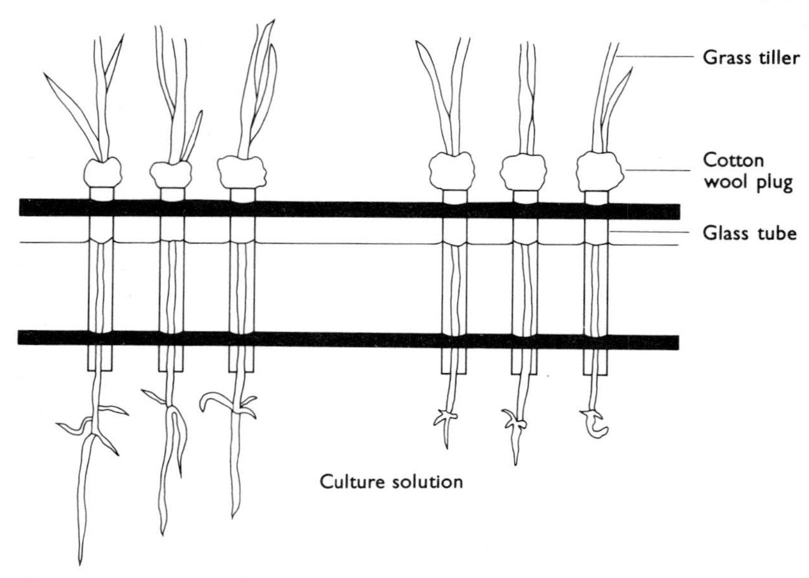

Fig. 5–1 Apparatus for measuring the metal tolerance of grasses in solution culture. Root growth of tolerant plants continues (left) but that of non-tolerant tillers ceases (right).

not confer tolerance to other metals. Mine spoil populations are tolerant of the metals present in toxic concentrations in the wastes on which they grow; they do not possess tolerance to other metals. For example, grasses tolerant of zinc ions in zinc mine spoil will not necessarily grow on copper mine spoil. Nevertheless, some plants exhibit multiple tolerance, being tolerant of more than one metal. Zinc and lead tolerances are frequently associated because zinc and lead often occur together in mine wastes.

Research on the genetics and evolution of heavy metal tolerance in grasses has proved that tolerant individuals can be selected from non-tolerant populations in a single step (WALLEY *et al.*, 1974). This was achieved by a screening method involving sowing seeds of *Agrostis tenuis* on toxic mixtures of metalliferous waste with soil. It is now clear that tolerance is controlled by many genes, which accounts for its high heritability and the observed fact that it is a continuously varying character. The genes are present in populations growing on non-metalliferous soils but there is such a wide range of gene combinations in *Agrostis tenuis* and other out-breeders that the appropriate gene combinations for a high degree of tolerance are very rare. Plants possessing these gene combinations can only be picked out by the powerful selection pressure exerted by mine spoil.

5.2.2 *Tolerance of other metals*

CHADWICK and SALT (1969) searched for tolerance of aluminium in *Agrostis tenuis* populations colonizing acid colliery waste which is known to contain high levels of aluminium in solution. They exposed rooted tillers to aluminium in solution culture, using the technique described for assaying heavy metal tolerance, and found that whereas root growth of tillers from a poor acid soil population was depressed by high aluminium, stimulation of root growth occurred in the acid spoil population. However, more recent work has failed to confirm the existence of aluminium tolerance in colliery spoil populations. Whilst aluminium may be present as a toxic factor in very acid spoils, the plants are affected more by nutrient deficiencies and direct effects of acidity.

Tolerance of magnesium and chromic ions has been shown to exist within populations of *Agrostis tenuis* native to serpentine soils but neither magnesium nor chromic toxicities are of importance on industrial wastes.

5.2.3 *Tolerance of low nutrient status*

Grasses exhibiting metal tolerance generally possess associated tolerances to other adverse growing conditions prevailing in the mine wastes on which they are found. JOWETT (1959) showed that a population of *Agrostis tenuis* from an old lead mine site was adapted to low levels of calcium and phosphate in the waste. Other studies have revealed that such ecotypes tend to be smaller in size than non-tolerant plants and are adapted to the exposure and drought conditions which occur on the tips.

5.3 Application of tolerance to reclamation

The importance of species selection has long been recognized in the planting of derelict and contaminated lands. It is often fairly simple to choose suitable species if information is available concerning the physico-chemical conditions of the substrates to be planted but, sometimes, species trials are necessary if unusual conditions prevail or the factors influencing growth are incompletely understood. In the United States, extensive species trials with trees and herbaceous plants have been conducted on strip mine spoils. In Denmark, trees and shrubs from all over the world have been tested for their tolerance and pioneer value on lignite spoil banks in the Danish 'Desert Arboretum'.

5.3.1 *Use of metal tolerant ecotypes*

SMITH and BRADSHAW (1970) have pointed out that metal tolerant grass populations provide a logical and economical way of overcoming most of the problems of metalliferous mine waste. They state that although the lack of nutrients can be overcome by the addition of fertilizers, there is no simple way of getting rid of toxicity unless tolerant ecotypes are planted (Table 20), particularly in combination with slow-acting fertilizers such as John Innes Base.

Table 20 Effects of nutrients on growth of normal and metal tolerant grasses on lead/zinc mine spoil. (After SMITH and BRADSHAW, 1970, *Nature*, **227**, 376–7.)

Species	*Grass tolerance*	*Dry weight per plant (mg)* Control	*NPK*	*John Innes base*
Festuca rubra	Tolerant	60	1370	3790
	Normal	30	110	240
Agrostis stolonifera	Tolerant	50	1510	1130
	Normal	10	260	130

GADGIL (1969) found that the metal tolerant mine populations of grasses discovered by Bradshaw in 1952 exhibited tolerance of metalliferous smelter wastes in the Lower Swansea Valley. The smelter wastes are more toxic than the corresponding mine spoils and normal grasses fail completely (Table 21). For tolerance to be an advantage, however, nutrients must be supplied to the smelter wastes.

Table 21 Growth of metal tolerant and non-tolerant *Agrostis tenuis* on smelter wastes.

Type of smelter waste	*Tolerance*	*Yields* ($g\ m^{-2}$) *Nil NPK*	*NPK fertilizer*
Zinc waste	Non-tolerant	0	0
	Zinc tolerant	0	50
Copper waste	Non-tolerant	0	4
	Copper tolerant	4	460

5.3.2 Ecological adjustment

One approach to effecting colonization has been to plant mixtures of species so that natural selection will adjust the species composition to meet the prevailing soil conditions. This process of ecological adjustment is well illustrated by the behaviour of mixed grass swards planted on blast furnace slag in the Lower Swansea Valley (Table 22).

Dactylis glomerata (cocksfoot) was the main species type in the seeds mixture and was dominant in the grass swards during the first two years. Thereafter, it declined markedly, being replaced by *Festuca rubra* which is

Table 22 Ecological adjustment of mixed swards planted on blast furnace slag.

Species type	*% species composition (dry weight)* Seeds mixture	2 years	4 years
Dactylis glomerata	50	58	18
Lolium spp.	26	4	0
Festuca rubra	10	9	64
Phleum pratense	8	16	13
Trifolium repens	6	11	3
Weed spp.	0	2	2

highly tolerant of alkaline soils and the poor, dry conditions which prevail in the slag.

Ecological adjustment can be utilized for effecting the colonization of metalliferous smelter wastes by metal tolerant ecotypes. At the present time, it is impractical to sow metal tolerant seed alone because of the following problems:

1. Metal tolerant seed is not commercially available in large quantities.
2. Tolerant ecotypes have inherently slow growth rates with the result that colonization occurs very slowly.
3. Tolerant seed multiplication is difficult because of selection against tolerance on uncontaminated soil.
4. Smelter wastes pose adverse seed-bed conditions and there is a high rate of failure at germination unless amendments are provided.

These difficulties can be overcome by sowing tolerant and normal seeds together, the former thinly and the latter thickly. The mixture is sown on minimally amended waste; thin layers of power station ash or organic materials can be used for amelioration. The non-tolerant seedlings establish quickly to form a non-persistent grass cover which is replaced in the second year by the vegetative spread of the tolerant plants (Fig. 5–2). Tolerant seedlings may also be produced which fill in the bare patches.

5·4 Ecology of wasteland communities

5·4·1 *Colliery waste*

HALL (1957) investigated the natural communities of colliery waste heaps and described the following successional plant communities:

1. Pioneer herb-grass community (on tips 10–15 years old).
2. Grass-herb (rich or average) communities (on tips 15–40 years old).
3. Grass-herb (acid) community (on tips of 40–80 years).
4. Grass-heath (on tips at least 60 years old).

Tips of up to 40 years old were found to be neutral or basic in soil

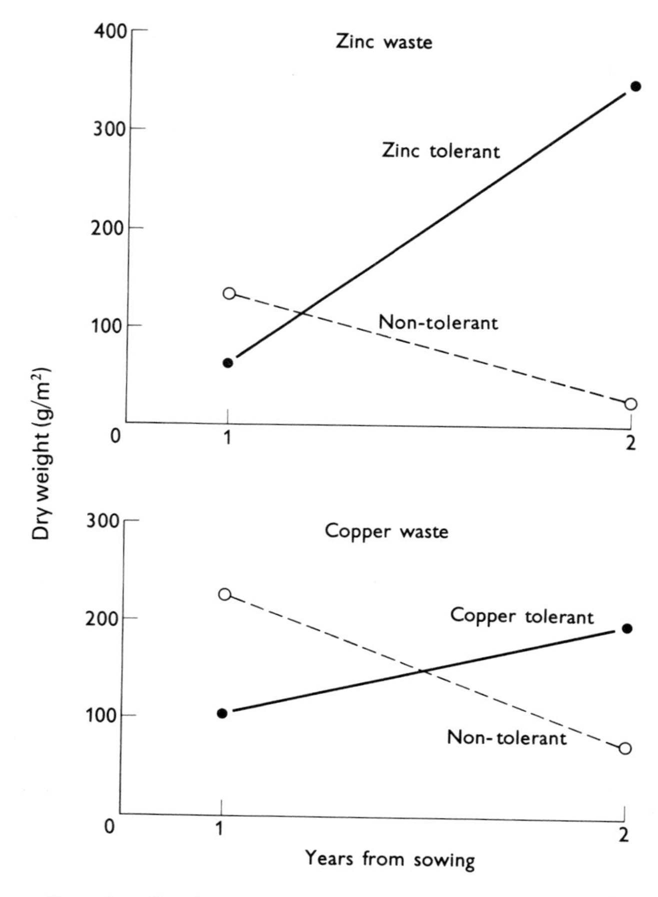

Fig. 5–2 Growth of tolerant and non-tolerant grass swards planted on minimally amended metalliferous smelter wastes.

reaction but later the spoil became very acid with an associated change in species composition. Thus a neutral or calcicolous flora was replaced by acid-heath grasses (*Deschampsia flexuosa*, *Agrostis tenuis* and *Nardus stricta*) with very few herb species. The most widespread and abundant wooded community was birch, which was often prominent on tips 30–40 years old. On some tips, hawthorn communities were common whereas on others the dominant species was oak. Gorse was the most frequent shrub species.

Hall's findings conflict with those of more recent workers in the North of England. In the Lancashire and Yorkshire coalfields, which he did not survey, spoil acidity develops extremely rapidly on most tips so that there is no succession from neutral or basic vegetation to acid heath.

5.4.2 *Alkali wastes and blast furnace slag*

These wastes often support extremely interesting vegetation consisting of open, calcicolous, herb-rich communities. The tip soils are very base-rich indeed and the low levels of major nutrients allow uncommon and interesting species to establish in the absence of competition. Immense populations of orchids have arisen on some alkali waste heaps in North-West England. On one site near Bolton (Fig. 5–3) the following species are abundant on the alkaline waste but are otherwise rare or absent in the natural habitats of the area:

Carlina vulgaris
Centaurium erythraea
Dactylorhiza fuchsii (see Fig. 5–4)
Dactylorhiza incarnata
Dactylorhiza purpurella
Erigeron acer
Euphrasia nemorosa var. *calcarea*
Gymnadenia conopsea
Linum catharticum
Orobanche minor
Sisyrinchium bermudiana

Of these species, *Carlina vulgaris, Centaurium erythraea, Erigeron acer, Euphrasia nemorosa* var. *calcarea* and *Linum catharticum* are typical calcicoles

Fig. 5–3 Alkali waste disposal site in the Croal-Irwell Valley near Bolton. The level tip of Leblanc Process waste in the centre of the picture has been undisturbed for 80 years and now supports calcicolous, herb-rich vegetation and orchids. Invasion by willow scrub is beginning on wet areas and hawthorns are appearing on well drained waste.

Fig. 5–4 *Dactylorhiza fuchsii* (common spotted orchid) in calcicolous vegetation on Leblanc waste site. Six species of orchid grow on the waste, some in colonies of several thousand plants.

of alkali waste tips, gas lime waste, limebeds, blast furnace slag, and the floors of limestone and chalk quarries.

Some of the most common and characteristic species of base-rich wastes include the following:

Achillea ptarmica	*Festuca rubra*
Achillea millefolium	*Hieracium* spp.
Agrostis stolonifera	*Lotus corniculatus*
Angelica sylvestris	*Pilosella officinalis*
Centaurea nigra	*Plantago lanceolata*
Crataegus monogyna	*Senecio jacobaea*
Dactylis glomerata	*Succisa pratensis*
Festuca ovina	*Tussilago farfara*

One lime waste bed in Cheshire (KELCEY, 1975) is of exceptional ecological interest and has been designated a nature reserve. Thirteen habitats have been recorded which support 252 species of flowering plants and ferns, several of which are not found elsewhere in the county.

Little is known concerning successional changes of the alkali waste communities although there is good evidence that developing hawthorn and willow scrub may become invasive and dramatically change the present flora of the sites.

6 Techniques of Wasteland Soil Analysis

6.1 Sampling

A common feature of wastes is that exposure to air, water, and fluctuating conditions of temperature and moisture causes dramatic chemical and physical changes to occur. Soluble chemical constituents are leached out of the superficial material but, at the same time, other soluble compounds may be released by the weathering of mineral fractions. Coarse fragments may disintegrate into finer particles whereas some wastes exhibit pozzolanic activity, i.e. they react with lime in the presence of moisture and undergo cementation as in power station ash. Salinity, pH and moisture holding capacity may change rapidly and to a considerable extent. Thus, the timing and conditions under which sampling is carried out are of paramount importance and should always be recorded.

Another characteristic of wastes is their great variability in structure and composition relative to most soils. Therefore, the methods of sample collection and the numbers of samples and subsamples obtained should allow for the nature and variability of the materials encountered, as well as the sizes of the areas under investigation.

6.1.1 *Number of samples*

Four composite samples should be regarded as an absolute minimum for assessing the substrate characteristics of any one site. For areas larger than 10 hectares or when different types of waste are obviously present, a greater number will be required.

Sampling points should be selected at random. Ideally, every sample should comprise ten or more sub-samples. These should be collected at random from within, say, a 25 metre radius of the main sampling point. The sub-samples should be bulked and well mixed to achieve homogenization.

6.1.2 *Depth of sampling*

Materials such as chemical wastes, power station ash and blast furnace slag should be sampled at a depth below the zone affected by leaching and weathering otherwise toxicity may be underestimated. Colliery spoil, on the other hand, is best sampled at or near the surface because toxicity in this waste is a consequence of exposure.

It is a good idea in most cases to collect both weathered and unweathered samples from the surface and at depth respectively; this will enable information on initial and potential toxicities to be obtained.

6.1.3 *Preparation of samples for analysis*

Samples must be dried in air as soon as possible after collection. They should be dried slowly at room temperature or not above 35°C; stronger heating must be avoided because it can cause an alteration in the chemical and physical nature, and therefore toxicity, of the samples. After drying, the wastes should be sieved so that particles of 2 mm diameter or less can be collected. Larger particles should be lightly crushed and any material which does not break up can be discarded; rock and other fragments have very little effect on soil chemical properties in most cases. Up to 50% or more of colliery spoil and blast furnace slag samples may be left behind by this sieving, whereas very little will remain when power station ash, some metal mine spoils, and certain chemical wastes are prepared for analysis.

6.2 pH determination

The pH of most types of waste can be determined in water suspension using a conventional pH meter fitted with a glass electrode. Colorimetric methods are often unreliable when applied to certain types of waste. Usually a 1 : 2.5 mixture of waste : water is appropriate but wherever possible the amount of water added should be kept to a minimum because the majority of wastes have a very low buffering capacity. It is best to find the minimum amount of water that has to be added in order to carry out the determinations and then employ that ratio of waste : water throughout the investigation.

6.3 Lime requirement

If pH levels are below 6.0 it is necessary to carry out lime requirement tests. It must be remembered, however, that lime requirement measurements can only determine the amount of lime needed to neutralize the initial acidity and there is no satisfactory method for determining potential acidity, as commonly occurs in colliery wastes.

The procedure is to incubate 10 g samples of waste with 20 cm^3 aliquots of Shoemaker buffer solution, the composition of which is detailed in Table 23. After incubation and shaking, the pH of the solution is measured. From this the lime requirement of the waste can be estimated by reference to a standard curve as shown in Fig. 6–1. The method works well for soils and some types of waste but recent studies have revealed that a number of precautions need to be taken for colliery shale. The most important modification is that the period of incubation should be extended from 30 minutes to 24 hours with constant shaking. Second, if the final waste/buffer solution pH is less than 5.5, the test must be

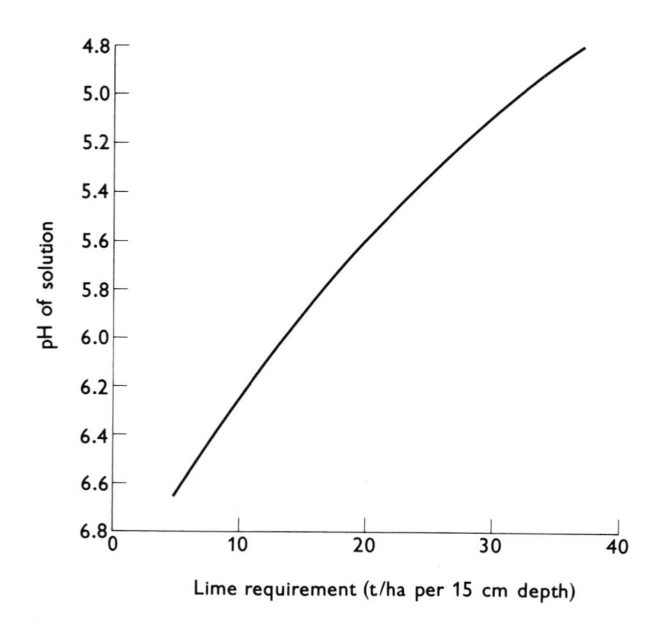

Fig. 6–1 Standard curve for lime requirement test. (After Shoemaker *et al.*, 1961, *Proc. Soil Sci. Soc. Am.*, **25**, 274.)

Table 23 Composition of buffer solution for lime requirement test. (After Shoemaker *et al.*, 1961, *Proc. Soil Sci. Soc. Am*, **25**, 274.)

p–nitrophenol	1.8 g
Triethanolamine	2.5 ml
Potassium chromate	3.0 g
Calcium acetate	2.0 g
Calcium chloride (dihydrate)	40.0 g

Dissolve in 800 cm³ distilled water,
Adjust pH to 7.5 with dilute HCl or NaOH,
Make up to 1000 cm³.

repeated using half the initial quantity of waste or twice the amount of buffer solution. For intensely acid types of colliery shale, the amount of buffer solution may need to be increased fourfold.

6.4 Phosphate status and fixation capacity

6.4.1 *Available phosphate*

There are numerous extraction methods for determining the amounts of plant-available phosphate in soils, many of which are only applicable to certain types of soil. Because wastes exhibit tremendous variation in chemical composition, it is impossible to recommend an extractant which is universally applicable. However, the most commonly used extractant is 0.5 N sodium bicarbonate solution at pH 8.5. This has been found to be suitable for colliery waste (FITTER and BRADSHAW, 1974) and gives reasonable results if applied to many other materials.

The procedure is to shake 2.5 g of waste with a 50 cm^3 aliquot of extractant for 1 hour. The suspension is then filtered or centrifuged and a 5 cm^3 aliquot is taken. The pH of this test solution is then set at 3.0 using 2.4 dinitrophenol solution as indicator; a buffer such as sodium sulphite should be employed to control this. 2.5 cm^3 of 0.02 M ammonium molybdate solution, 9.8 N for sulphuric acid, is then added and the mixture diluted to 25 cm^3. The molybdate complex formed is then chemically reduced by addition of 0.1 cm^3 of M stannous chloride solution in 1.25 N hydrochloric acid. Finally, the intensity of the blue colour of the resulting phospho-molybdate-tin complex is determined colorimetrically at 660 nm (see Table 24).

Table 24 Interpretation of phosphorus and potassium analytical results.

Available P *or* K *(ppm)* P	K	*Fertilizer requirement*
0– 5	0– 40	High
5–10	40– 80	Moderate
10–20	80–100	Low
20+	100+	None

6.4.2 *Phosphate fixation capacity*

As described in Chapter 2, fixation of phosphate is of common occurrence in colliery spoil. Therefore, determinations of available phosphate may be of little or no value in predicting fertilizer requirements. Fixation capacity may be estimated by incubating 2.5 g samples of waste in 50 cm^3 aliquots of 0.01 M calcium chloride solution containing 20 ppm P as calcium tetrahydrogen diorthophosphate ($Ca(H_2PO_4)_2 2H_2O$) as recommended by HESSE (1971). The samples should be shaken for 24 hours, filtered, and the residual phosphate in solution determined as described in the previous section. From the change in phosphate concentration in solution, it is possible to estimate the fixation capacity of the waste.

6.5 Potassium status

Potassium is extracted with 0.5 N ammonium acetate solution at pH 7.0 as recommended by DOUBLEDAY (1971). A 10 g sample of waste is treated with 50 cm^3 of extractant and shaken for 30 minutes. After filtration, potassium in solution can be determined by flame emission photometry (see Table 24).

6.6 Nitrogen status

Wastes are generally extremely low in both total and available nitrogen. Colliery shale is an exception in that fossilized nitrogen may be present, but this rarely or never becomes available to plants. It is recommended that nitrogen determinations be omitted.

6.7 Salinity

Salinity is measured in terms of the electrical conductance of a saturated moisture extract. This is prepared by adding deionized water to about 100 cm^3 of waste until puddling occurs. At this stage, it should be impossible to remove water from the waste except by vacuum filtration. DOUBLEDAY (1971) recommends that the samples should be left for a period of two hours because some salts dissolve very slowly and imbibition of moisture may proceed for some time. The saturated paste is then extracted in a Buchner funnel.

After suction removal of the extract, its conductance is determined by means of a Wheatstone Bridge conductivity meter (see Table 25).

Table 25 Interpretation of conductance results.

Conductance (mmhos/cm at 25°C)	*Degree of salinity*	*Effect on plants*
0–2	Non-saline	None
2–5	Slightly saline	Growth reduction of susceptible plants only
5–10	Moderately saline	Growth reduction of most plants
10–15	Highly saline	Serious growth reduction of all plants
15+	Intensely saline	Growth extremely poor or nil

6.8 Trace element and minor nutrient determinations

Determinations of trace element levels in wasteland soils reveal excesses rather than deficiencies in most cases.

It has been pointed out in Chapter 4 that conventional types of soil extractant like ammonium acetate and acetic acid tend to give over-estimates of the plant-available concentrations of trace metals such as zinc, copper and lead in wastes. Simple water extraction techniques are therefore recommended for first investigations. If organic matter or clays are present, it is advised that metal determinations be conducted on normal ammonium acetate extracts as well.

Colorimetric methods are available for the analysis of the majority of metal cations in solution but they are time-consuming and often difficult to perform. In recent years, atomic absorption spectrophotometry has been widely employed because of its simplicity, convenience, and freedom from interference effects caused by other ions in solution. Flame emission photometry is normally used for potassium and sodium determinations.

Titration methods are applicable for analyses of certain anions in solution. Soluble carbonates, hydroxides and bicarbonates can be estimated by titration with dilute sulphuric acid. Similarly, chlorides may be determined by titration with silver nitrate. Sulphates are more difficult but turbimetric methods have been successfully employed by some workers.

Most textbooks of soil chemical analysis describe methods of determining the presence and amounts of cations and anions in solution.

6.9 Analysis of plant growth responses

Scientific experiments on the growth of plants on wastes can be carried out using very simple equipment and readily available chemicals or fertilizers. Plastic plant pots are ideal for experiments conducted under glass or outside and their use permits a large number of treatments to be investigated with the minimum of effort and time. Trials on the wastes *in situ* must involve larger experimental units or plots because of edge effects, site variation and possible disturbance by animals and man.

The design of experiments is of crucial importance because of the problem of variability in the physical structure and chemical composition of most wastes. For experiments to be of practical significance, the treatments should be repeated several times so as to cover the range of materials encountered on site. Every experiment, therefore, must contain replicated treatments and randomization of pots or plots as shown in the trial area illustrated in Fig. 6–2. The randomized block layout (Fig. 6–4) is the most generally applicable design but there are several useful alternative layouts.

Fig. 6–2 Experimental grass plots on a levelled blast furnace slag tip at Ulverston in Cumbria. Seeds of tolerant grass species such as *Festuca rubra* were sown direct into waste which was treated with various types and rates of nitrogen and phosphate fertilizers. The main

Lolium perenne (perennial ryegrass) is a useful experimental test plant because it grows quickly and is often planted in reclamation work. Quantitative information on the effects of treatments can be obtained by cutting the grass and measuring its dry weight. The effects of nutrients, liming materials and other toxicity ameliorants can be evaluated in this way.

6.9.1 *Complete randomization and randomized blocks*

In most experiments, at least four treatments will be under investigation. For example, if the effects of nitrogen and phosphate fertilizer applications are tested alone and in combination, the four treatments will be as follows:

A No nutrients
B Nitrogen only
C Phosphate only
D Nitrogen and phosphate

If there are four replicates in the experiments, there will be a total of 16 pots or plots. This is often called a factorial experiment replicated four times, the two factors involved being nitrogen fertilization and phosphate fertilization.

The simplest experimental design involves complete randomization of the treatments and replicates in a single block (Fig. 6–3). This design is often acceptable for glasshouse experiments conducted in pots but is a rather crude design for field trials where site variation can be very large.

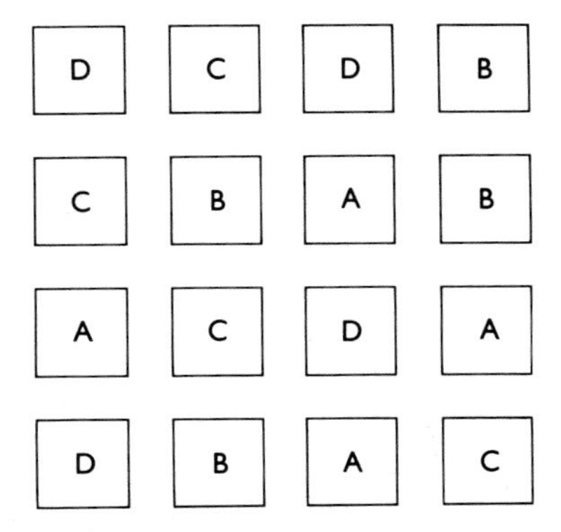

Fig. 6–3 Layout of a completely randomized experiment.

The randomized block layout overcomes most of the problems caused by differences in site materials and other variables. Every treatment is represented once in each block (Fig. 6–4), their arrangement within the blocks being at random. The blocks themselves may be close together or

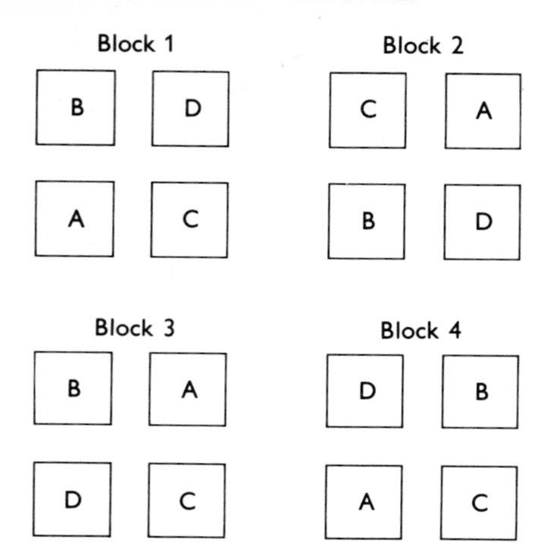

Fig. 6–4 Layout of a simple randomized block experiment.

well separated on different parts of the tip site being studied. Even if one block is on relatively innocuous waste and another on highly toxic material, the experiment does not lose its precision. Indeed, the major advantage of this layout is that the results are more applicable to the site as a whole than if the experiment had been conducted in a single block as described previously.

6.9.2 *Latin Squares*

Another modification of the completely randomized layout is the Latin Square (Fig. 6–5). Here, the plots are arranged in one block but, instead of being completely at random, every row and every column of plots in the square contains every one of the four treatments A, B, C and D which

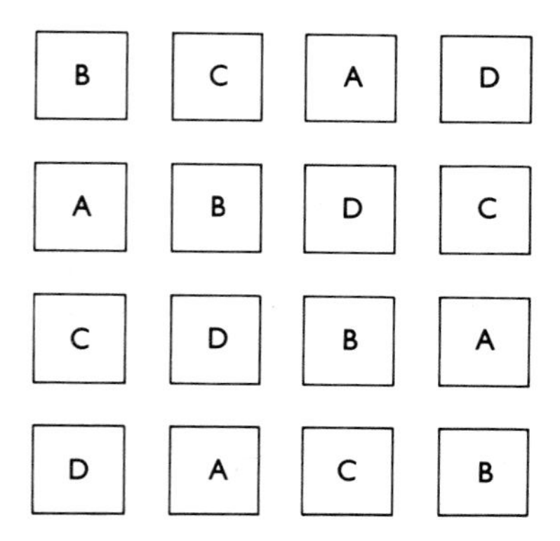

Fig. 6–5 Layout of a simple Latin Square experiment.

are distributed at random. The advantage of the Latin Square is that differences in substrate fertility or toxicity both across and down the block do not reduce the precision of the experiment.

6.9.3 *Split-plots*

This technique of design can be applied to any of the three previous types of layout. In its simplest form, every plot is split in half, one half-plot being treated and the other left untreated as control (Fig. 6–6). It is particularly useful where maintenance treatments such as fertilizer or lime top-dressings are to be investigated during the course of an experiment. It will be noticed from Fig. 6–6 that the maintenance treatment may be imposed on either of the two split-plots on a random basis.

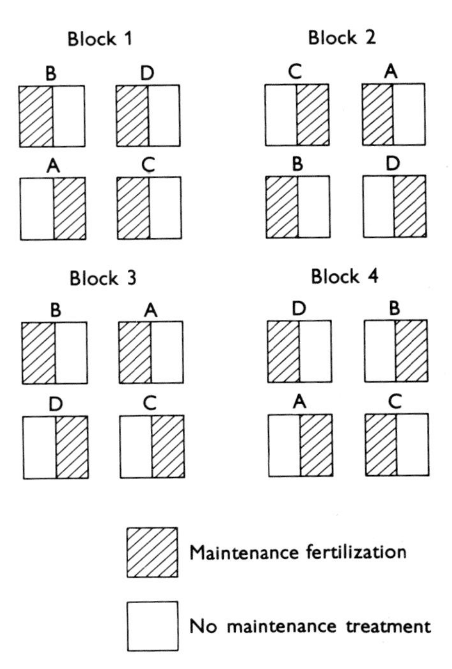

Fig. 6–6 Design of a split-plot experiment based on a randomized block layout.

6.9.4 *Statistical analysis of results*

The dry weight data obtained from properly designed and replicated experiments should be assessed by analysis of variance in order to determine whether any observed effects of treatments are significant or not. When treatment effects are statistically significant, the values of means can be compared by Duncan's New Multiple Range Test. Full details and guidance on the statistical analysis of experimental results may be obtained from two earlier booklets in the 'Studies in Biology' series:

No. 23 *Investigation by Experiment* (O. V. S. HEATH)
No. 43 *Introductory Statistics for Biology* (R. E. PARKER)

Further Reading

CHADWICK, M. J. and GOODMAN, G. T., eds. (1975). *The Ecology of Resource Degradation and Renewal*. Blackwell Scientific Publications, Oxford.

GOODMAN, G. T., EDWARDS, R. W. and LAMBERT, J. M., eds. (1965). *Ecology and the Industrial Society*. Blackwell Scientific Publications, Oxford.

HUTNIK, R. J. and DAVIS, G., eds. (1973). *Ecology and Reclamation of Devastated Land*, Vols. 1 and 2. Gordon and Breach, New York and London.

UNIVERSITY OF NEWCASTLE UPON TYNE (1971 and 1972). *Landscape Reclamation*, Vols. 1 and 2. IPC Science and Technology Press Ltd., Guildford.

References

BRADSHAW, A. D. (1952). *Nature*, **169**, 1098.

BRADSHAW, A. D. (1970). *Trans. Bot. Soc. Edinb.*, **41**, 71–84.

BREEZE, V. (1973). *J. Appl. Ecol.*, **10**, 513–25.

CARUCCIO, F. T. (1973). In *Ecology and Reclamation of Devastated Land*, Vol. 1, eds. HUTNIK, R. J. and DAVIS, G. Gordon and Breach, New York and London.

CHADWICK, M. J. (1973). In *Ecology and Reclamation of Devastated Land*, Vol. 1, eds. HUTNIK, R. J. and DAVIS, G. Gordon and Breach, New York and London.

CHADWICK, M. J., CORNWELL, S. M. and PALMER, M. E. (1969). *Nature*, **222**, 161.

CHADWICK, M. J. and SALT, J. K. (1969). *Nature*, **224**, 186.

CRESSWELL, C. F. (1973). In *Ecology and Reclamation of Devastated Land*, Vol. 2, eds. HUTNIK, R. J. and DAVIS, G. Gordon and Breach, New York and London.

DAVISON, A. and JEFFERIES, B. J. (1966). *Nature*, **210**, 649–50.

DEKOCK, P. C. (1956). *Ann. Bot.*, **20**, 133–41.

DOUBLEDAY, G. P. (1971). In *Landscape Reclamation*, Vol. 1. IPC Science and Technology Press Ltd., Guildford.

DOUBLEDAY, G. P. (1972). In *Landscape Reclamation*, Vol. 2. IPC Science and Technology Press Ltd., Guildford.

FITTER, A. H. and BRADSHAW, A. D. (1974). *J. Appl. Ecol.*, **11**, 597–608.

GADGIL, R. L. (1969). *J. Appl. Ecol.*, **6**, 247–59.

GEMMELL, R. P. (1972) *Nature*, **240**, 569–71.

GEMMELL, R. P. (1973). *Environ. Pollut.*, **5**, 181–97.

GEMMELL, R. P. (1974). *Environ. Pollut.*, **6**, 31–7.

GEMMELL, R. P. (1974). *Nature*, **247**, 199–200.

GEMMELL, R. P. (1975). *Environ. Pollut.*, **8**, 35–44.

GOODMAN, G. T., PITCAIRN, C. E. R. and GEMMELL, R. P. (1973). In *Ecology and Reclamation of Devastated Land*, Vol. 2, eds. HUTNIK, R. J. and DAVIS, G. Gordon and Breach, New York and London.

GREGORY, R. P. G. and BRADSHAW, A. D. (1965). *New Phytol.*, **64**, 131–43.

HALL, I. G. (1957). *J. Ecol.*, **45**, 689–720.

HESSE, P. R. (1971). *A Textbook of Soil Chemical Analysis*. John Murray, London.

HEWITT, E. J. (1948). *Nature*, **161**, 489–90.

HILL, R. D. (1973). In *Ecology and Reclamation of Devastated Land*, Vol. 2, eds. HUTNIK, R. J. and DAVIS, G. Gordon and Breach, New York and London.

HODGSON, D. R., HOLLIDAY, R. and COPE, F. (1963). *J. Agric. Sci.*, **61**, 299–308.

HODGSON, D. R. and TOWNSEND, W. N. (1973). In *Ecology and Reclamation of Devastated Land*, Vol. 2, eds. HUTNIK, R. J. and DAVIS, G. Gordon and Breach, New York and London.

HOLLIDAY, R., HODGSON, D. R., TOWNSEND, W. N. and WOOD, J. W. (1958). *Nature*, **176**, 1079–80.

JOWETT, D. (1958). *Nature*, **182**, 816.

JOWETT, D. (1959). *Nature*, **184**, 43.

KELCEY, J. G. (1975). *Environ. Conserv.*, **2**, 99–108.

KNABE, W. (1973). In *Ecology and Reclamation of Devastated Land*, Vol. 2, eds. HUTNIK, R. J. and DAVIS, G. Gordon and Breach, New York and London.

LE ROUX, N. W. (1969). *New Scientist*, **25**, September 1969, 14–16.

NEWTON, A. (1971). *Flora of Cheshire*. Cheshire Community Council, Chester.

REES, W. J. and SIDRAK, G. H. (1956). *Plant and Soil*, **8**, 141–59.

REPP, G. (1973). In *Ecology and Reclamation of Devastated Land*, Vol. 1, eds. HUTNIK, R. J. and DAVIS, G. Gordon and Breach, New York and London.

SCHMEHL, W. R. and McCASLIN, B. D. (1973). In *Ecology and Reclamation of Devastated Land*, Vol. 1, eds. HUTNIK, R. J. and DAVIS, G. Gordon and Breach, New York and London.

SMITH, R. A. H. and BRADSHAW, A. D. (1970). *Nature*, **227**, 376–7.

STREET, H. E. and GOODMAN, G. T. (1967). In *The Lower Swansea Valley Project*, ed. HILTON, K. J. Longmans Green, London.

TOWNSEND, W. N. and HODGSON, D. R. (1973). In *Ecology and Reclamation of Devastated Land*, Vol. 2, eds. HUTNIK, R. J. and DAVIS, G. Gordon and Breach, New York and London.

WALLEY, K. A., KHAN, M. S. I. and BRADSHAW, A. D. (1974). *Heredity*, **32**, 309–19.

WILKINS, D. A. (1957). *Nature*, **180**, 37.

122992

LINKED